Springer Undergraduate Mathematics Series

For other titles published in this series, go to
www.springer.com/series/3423

Norman L. Biggs

Codes: An Introduction to Information Communication and Cryptography

Norman L. Biggs
Department of Mathematics
London School of Economics
Houghton Street
London WC2A 2AE, UK

Springer Undergraduate Mathematics Series ISSN 1615-2085
ISBN: 978-1-84800-272-2 e-ISBN: 978-1-84800-273-9
DOI: 10.1007/978-1-84800-273-9

British Library Cataloguing in Publication Data
A catalogue record for this book is available from the British Library

Library of Congress Control Number: 2008930146

Mathematics Subject Classification (2000): 94A, 94B, 11T71

Printed on acid-free paper

9 8 7 6 5 4 3 2 1

Springer Science+Business Media
springer.com

Preface

Many people do not realise that mathematics provides the foundation for the devices we use to handle information in the modern world. Most of those who do know probably think that the parts of mathematics involved are quite 'classical', such as Fourier analysis and differential equations. In fact, a great deal of the mathematical background is part of what used to be called 'pure' mathematics, indicating that it was created in order to deal with problems that originated within mathematics itself. It has taken many years for mathematicians to come to terms with this situation, and some of them are still not entirely happy about it.

This book is an integrated introduction to Coding. By this I mean replacing symbolic information, such as a sequence of bits or a message written in a natural language, by another message using (possibly) different symbols. There are three main reasons for doing this: Economy (data compression), Reliability (correction of errors), and Security (cryptography).

I have tried to cover each of these three areas in sufficient depth so that the reader can grasp the basic problems and go on to more advanced study. The mathematical theory is introduced in a way that enables the basic problems to be stated carefully, but without unnecessary abstraction. The prerequisites (sets and functions, matrices, finite probability) should be familiar to anyone who has taken a standard course in mathematical methods or discrete mathematics. A course in elementary abstract algebra and/or number theory would be helpful, but the book contains the essential facts, and readers without this background should be able to understand what is going on.

There are a few places where reference is made to computer algebra systems. I have tried to avoid making this a prerequisite, but students who have access to such a system will find it helpful. In particular, there are occasional specific references to MAPLE™ (release 10), by Maplesoft, a division of Waterloo Maple Inc., Waterloo, Canada.

The book has been developed from a course of twenty lectures on Information, Communication, and Cryptography given for the MSc in Applicable Mathematics at the London School of Economics. I should like to thank all those students who have contributed to the development of the course materials, in particular those who have written dissertations in this area: Rajni Kanda, Ovijit Paul, Arunduti Dutta-Roy, Ana de Corbavia-Perisic, Raminder Ruprai, James Rees, Elisabeth Biell, Anisa Bhatt, Timothy Morill, Shivam Kumar, and Carey Chua. I owe a special debt to Raminder Ruprai, who worked through all the exercises and helped to sort out many mistakes and obscurities.

Finally, I am grateful to Aaron Wilson, who helped to produce the diagrams, and especially to Karen Borthwick, who has been very helpful and supportive on behalf of the publishers.

Norman Biggs
January 2008

Contents

Preface v

1. Coding and its uses 1
1.1 Messages 1
1.2 Coding 3
1.3 Basic definitions 4
1.4 Coding for economy 7
1.5 Coding for reliability 8
1.6 Coding for security 9

2. Prefix-free codes 13
2.1 The decoding problem 13
2.2 Representing codes by trees 16
2.3 The Kraft-McMillan number 18
2.4 Unique decodability implies $K \leq 1$ 21
2.5 Proof of the Counting Principle 24

3. Economical coding 27
3.1 The concept of a source 27
3.2 The optimization problem 30
3.3 Entropy 32
3.4 Entropy, uncertainty, and information 34
3.5 Optimal codes – the fundamental theorems 38
3.6 Huffman's rule 40
3.7 Optimality of Huffman codes 44

4. Data compression 47
4.1 Coding in blocks 47
4.2 Distributions on product sets 49
4.3 Stationary sources 52
4.4 Coding a stationary source 55
4.5 Algorithms for data compression 58
4.6 Using numbers as codewords 59
4.7 Arithmetic coding 62
4.8 The properties of arithmetic coding 65
4.9 Coding with a dynamic dictionary 67

5. Noisy channels 73
5.1 The definition of a channel 73
5.2 Transmitting a source through a channel 76
5.3 Conditional entropy 78
5.4 The capacity of a channel 81
5.5 Calculating the capacity of a channel 83

6. The problem of reliable communication 89
6.1 Communication using a noisy channel 89
6.2 The extended BSC 94
6.3 Decision rules 96
6.4 Error correction 100
6.5 The packing bound 102

7. The noisy coding theorems 107
7.1 The probability of a mistake 107
7.2 Coding at a given rate 111
7.3 Transmission using the extended BSC 113
7.4 The rate should not exceed the capacity 117
7.5 Shannon's theorem 119
7.6 Proof of Fano's inequality 120

8. Linear codes 123
8.1 Introduction to linear codes 123
8.2 Construction of linear codes using matrices 126
8.3 The check matrix of a linear code 128
8.4 Constructing 1-error-correcting codes 131
8.5 The decoding problem 135

9. Algebraic coding theory ... 141
9.1 Hamming codes ... 141
9.2 Cyclic codes ... 145
9.3 Classification and properties of cyclic codes ... 149
9.4 Codes that can correct more than one error ... 153
9.5 Definition of a family of BCH codes ... 155
9.6 Properties of the BCH codes ... 158

10. Coding natural languages ... 163
10.1 Natural languages as sources ... 163
10.2 The uncertainty of `english` ... 165
10.3 Redundancy and meaning ... 168
10.4 Introduction to cryptography ... 170
10.5 Frequency analysis ... 174

11. The development of cryptography ... 179
11.1 Symmetric key cryptosystems ... 179
11.2 Poly-alphabetic encryption ... 180
11.3 The Playfair system ... 183
11.4 Mathematical algorithms in cryptography ... 185
11.5 Methods of attack ... 187

12. Cryptography in theory and practice ... 191
12.1 Encryption in terms of a channel ... 191
12.2 Perfect secrecy ... 195
12.3 The one-time pad ... 197
12.4 Iterative methods ... 198
12.5 Encryption standards ... 201
12.6 The key distribution problem ... 203

13. The RSA cryptosystem ... 207
13.1 A new approach to cryptography ... 207
13.2 Outline of the RSA system ... 209
13.3 Feasibility of RSA ... 212
13.4 Correctness of RSA ... 215
13.5 Confidentiality of RSA ... 217

14. Cryptography and calculation ... 221
14.1 The scope of cryptography ... 221
14.2 Hashing ... 222
14.3 Calculations in the field $\mathbb{F}_p$ ... 224
14.4 The discrete logarithm ... 226

14.5 The ElGamal cryptosystem 228
14.6 The Diffie-Hellman key distribution system 230
14.7 Signature schemes 232

15. Elliptic curve cryptography 237
15.1 Calculations in finite groups 237
15.2 The general ElGamal cryptosystem 239
15.3 Elliptic curves 241
15.4 The group of an elliptic curve 245
15.5 Improving the efficiency of exponentiation 248
15.6 A final word 250

Answers to odd-numbered exercises 255

Index 271

1
Coding and its uses

1.1 Messages

The first task is to set up a simple mathematical model of a message. We do this by looking at some examples and extracting some common features from them.

Example 1.1

Many messages are written in a natural language, such as English. These messages contain symbols, and the symbols form words, which in turn form sentences, such as this one. The messages may be sent from one person to another in several ways: in the form of a handwritten note or an email, for example. A text message is essentially the same, but it is often expressed in an unnatural language.

Example 1.2

Devices such as scanners and digital cameras produce messages in the form of electronic impulses. These messages may be sent from one device to another by wires or optic fibres, or by radio waves.

Formal definitions based on these examples will be given in Section 1.3. For the time being, we shall think of a message as a sequence of symbols, noting

N. L. Biggs, *An Introduction to Information Communication and Cryptography*,
DOI: 10.1007/978-1-84800-273-9_1,

that the order of the symbols is clearly important.

The function of a message is to convey information from a sender to a receiver. In order to do this successfully, the sender and receiver must agree to use the same set of symbols. This set is called an *alphabet*.

Example 1.3

We denote by $\mathbb{A}$ the alphabet which has 27 symbols, the letters A, B, C, ..., Z, and a 'space', which we denote by ⊔. We shall often use the alphabet $\mathbb{A}$ to represent messages written in English. This is convenient for the sake of exposition, but obviously some features are ignored. Thus we ignore the distinction between upper and lower case letters, and we omit punctuation marks. Of course, there may be some loss in reducing an English message into a string of symbols in this alphabet. For example the text

The word 'hopefully' is often misused.

is reduced to the following message in $\mathbb{A}$.

THE⊔WORD⊔HOPEFULLY⊔IS⊔OFTEN⊔MISUSED

Example 1.4

The alphabet $\mathbb{B}$ has 2 symbols, 0 and 1, which are called *binary digits* or *bits*. Because the bits 0 and 1 can be implemented electronically as the states OFF and ON, this is the underlying alphabet for all modern applications. In practice, the bits are often combined into larger groups, such as '32-bit words'. But any message that is transmitted electronically, whether it originates as an email from me or as an image from a satellite orbiting the earth, is essentially a sequence of bits.

EXERCISES

1.1. The following messages have been translated from 'proper English' into the alphabet $\mathbb{A}$. Write down the original messages and comment upon any ambiguity or loss of meaning that has occurred.

CANINE⊔HAS⊔SIX⊔LETTERS⊔AND⊔ENDS⊔IN⊔NINE

ITS⊔HOT⊔SAID⊔ROBERT⊔BROWNING

1.2. A 32-bit word is a sequence of 32 symbols from the alphabet $\mathbb{B}$. How many different 32-bit words are there? If my printer can print one every second, how many years (approximately) will it take to print them all?

1.3. In the period 1967-86 the ASCII alphabet was widely used as a standard for electronic communication. It has 128 symbols, 95 of which were printable. In this book we have already used some symbols that were not in the ASCII alphabet. Which ones? [ASCII is an abbreviation for *American Standard Code for Information Interchange*, and is pronounced 'askey'. The ASCII alphabet is now part of a much more comprehensive system known as Unicode.]

1.4. Not all natural languages use 26 letters. How many letters are there in (i) the modern Greek alphabet and (ii) the Russian Cyrillic alphabet?

1.2 Coding

Roughly speaking, coding is a rule for replacing one message by another message. The second message may or may not use the same alphabet as the first.

Example 1.5

A simple rule for coding messages in the 27-symbol alphabet $\mathbb{A}$ using the same alphabet is: *write each word backwards.* So the message

SEE␣YOU␣TOMORROW becomes EES␣UOY␣WORROMOT .

Example 1.6

A rule for coding messages in $\mathbb{A}$ using the binary alphabet $\mathbb{B}$ is: *replace vowels by* 0, *replace consonants by* 1, *and ignore the spaces.* With this rule

SEE␣YOU␣TOMORROW becomes 10010010101101 .

These two examples are very artificial, and the rules are of limited value. For greater realism and utility we must look at the purposes for which coding is used, and evaluate proposed coding rules in that context.

There are three major reasons for coding a message.

ECONOMY In many situations it is necessary or desirable to use an alphabet smaller than those that occur in natural languages. It may also be desirable to make the message itself smaller: in recent times this has led to the development of techniques for *Data Compression.*

RELIABILITY Messages may be altered by 'noise' in the process of transmission. Thus there is a need for codes that allow for *Error Correction.*

SECURITY Some messages are sent with the requirement that only the right person can understand them. Historically, secrecy was needed mainly in diplomatic and military communications, but nowadays it plays an important part in everyday commercial transactions. This area of coding is known as *Cryptography.*

EXERCISES

1.5. The following messages are coded versions of meaningful English sentences. Explain the coding rules used and find the original messages.

7 15 15 4 27 12 21 3 11

00111 01111 01111 00100 11011 01100 10101 00011 01011

1.6. Explain formally (as if you were writing a computer program) the coding rule *write each word backwards.* [You must explain how to convert a sequence of symbols such as `TODAY␣IS␣MONDAY` into `YADOT␣SI␣YADNOM`.]

1.3 Basic definitions

We are now ready to make some proper definitions.

Definition 1.7 (Alphabet)

An *alphabet* is a finite set S; we shall refer to the members of S as *symbols*.

Definition 1.8 (Message, string, word)

A *message* in the alphabet S is a finite sequence of members of S:

$$x_1x_2\cdots x_n \qquad (x_i \in S,\ 1 \leq i \leq n).$$

A message is often referred to as a *string* of members of S, or a *word* in S. The number n is called the *length* of the message, string, or word.

The set of all strings of length n is denoted by S^n. For example, when $S = \mathbb{B}$ and $n = 3$, the set $\mathbb{B}^3$ consists of the strings

$$000 \quad 001 \quad 010 \quad 011 \quad 100 \quad 101 \quad 110 \quad 111 \quad .$$

The set of all strings in S is denoted by S^*:

$$S^* = S^0 \cup S^1 \cup S^2 \cup \cdots \quad .$$

Note that S^0 consists of the string with length zero; in other words, the string with no symbols. We include it in the definition because sometimes it is convenient to use it.

Definition 1.9 (Code, codeword)

Let S and T be alphabets. A *code* c for S using T is an injective function $c : S \to T^*$. For each symbol $s \in S$ the string $c(s) \in T^*$ is called the *codeword* for s. The set of all codewords,

$$C = \{c(s) \mid s \in S\},$$

is also referred to as the code. When $|T| = 2$ the code is said to be *binary*, when $|T| = 3$ it is *ternary*, and in general when $|T| = b$, it is *b-ary*.

For example, let $S = \{x, y, z\}$, $T = \mathbb{B}$, and define

$$c(x) = 0, \qquad c(y) = 10, \qquad c(z) = 11.$$

This is a binary code, and the set of codewords is $C = \{0, 10, 11\}$.

According to the definition, a code c assigns to each *symbol* in S a *string of symbols* in T. The strings may vary in length. For example, suppose we are trying to construct a code for the 27-symbol English alphabet $\mathbb{A}$ using the binary alphabet $\mathbb{B}$. We might begin by choosing codewords of length 4, as follows:

$$\mathtt{A} \mapsto 0000 \quad \mathtt{B} \mapsto 0001 \quad \mathtt{C} \mapsto 0010 \quad \ldots \quad .$$

Now, the definition requires c to be an injective function or (as we usually say) an *injection*. This is the mathematical form of the very reasonable requirement that c does not assign the same codeword to two different symbols. In other words, if $c(s) = c(s')$ then $s = s'$. Clearly, there are only 16 strings of length 4 in $\mathbb{B}$, so the 27 symbols in $\mathbb{A}$ cannot all be assigned different ones.

Thus far we have considered only the coding of individual symbols. The extension to messages (strings of symbols) is clear.

Definition 1.10 (Concatenation)

If $c : S \to T^*$ is a code, we extend c to S^* as follows. Given a string $x_1 x_2 \cdots x_n$ in S^*, define

$$c(x_1 x_2 \cdots x_n) = c(x_1)c(x_2)\cdots c(x_n).$$

This process is known as *concatenation*. Note that we denote the extended function $S^* \to T^*$ by the same letter c.

It is not always possible to recover the original string uniquely from the coded version. For example, let $S = \{x, y, z\}$, and define $c : S \to \mathbb{B}^*$ by

$$x \mapsto 0, \quad y \mapsto 01, \quad z \mapsto 10.$$

Suppose we are given the string 010100 which, we are told, is the result of coding a string in S^* using c. By trial and error we find two possibilities (at least):

$$xzzx \mapsto 0\,10\,10\,0\,, \qquad yyxx \mapsto 01\,01\,0\,0\,.$$

Clearly, this situation is to be avoided, if possible.

Definition 1.11 (Uniquely decodable)

The code $c : S \to T^*$ is *uniquely decodable* (or *UD* for short) if the extended function $c : S^* \to T^*$ is an injection. This means that any string in T^* corresponds to at most one message in S^*.

In Chapter 2 we shall explain how the UD property can be guaranteed by a simple construction.

EXERCISES

1.7. A binary code is defined by the rule

$$s_1 \mapsto 10, \quad s_2 \mapsto 010, \quad s_3 \mapsto 100.$$

Show by means of an example that this code is not uniquely decodable.

1.8. Suppose the code $c : S \to T^*$ is such that every codeword $c(s)$ has the same length n. Is this code uniquely decodable?

1.9. Express the coding rules used in Exercise 1.5 as functions $c : S \to T^*$, for suitable alphabets S and T.

1.4 Coding for economy

When the electric telegraph was first introduced, it could transmit only simple electrical impulses. Thus, in order to send messages in a natural language it was necessary to code them into an alphabet with very few symbols. A suitable code was invented by Samuel Morse (1791-1872).

The *Morse Code* uses an alphabet of three symbols: $\{\bullet, -, \odot\}$. The $\bullet$ (*dot*, pronounced *di*) is a short impulse, the $-$ (*dash*, pronounced *dah*) is a long impulse, and the $\odot$ is a pause. Every codeword comprises dots and dashes, ending with a pause. (Strictly speaking, there are also symbols for the shorter pause that separates the dots and dashes within a codeword, and for the longer pause at the end of a message word, but we shall ignore them for the sake of simplicity.) Here are the codewords for $A, B, C, D, E, F, X, Y, Z$.

$$A \mapsto \bullet - \odot \qquad B \mapsto - \bullet \bullet \bullet \odot \qquad C \mapsto - \bullet - \bullet \odot$$
$$D \mapsto - \bullet \bullet \odot \qquad E \mapsto \bullet \odot \qquad F \mapsto \bullet \bullet - \bullet \odot$$

$$X \mapsto - - \bullet \bullet - \odot \quad Y \mapsto - \bullet - - \odot \quad Z \mapsto - - \bullet \bullet \odot$$

In Chapters 2, 3, and 4 we shall look at the basic theory of economical coding and explain how it can be applied to the compression of data. This subject has become very important, because huge amounts of data are now being generated and transmitted electronically.

EXERCISES

1.10. Search the internet to find the standard version of Morse Code, known as the International Morse Code. If this code is defined formally as a function $S \to T^*$, what are the alphabets S and T?

1.11. Decode the following Morse messages:

$$\bullet \bullet \bullet \odot - - - \odot \bullet \bullet \bullet \odot \quad ;$$

$$- - \odot \bullet - \odot - \bullet - - \odot - \bullet \bullet \odot \bullet - \odot - \bullet - - \odot \quad .$$

1.12. Suppose we try to use a version of Morse code without the symbol $\odot$ that indicates the end of each codeword. What is the code for BAD? Find another English word with the same code, showing that this is not a uniquely decodable code.

1.13. The *semaphore* code enables messages to be exchanged between people who can see each other. Each person has two flags, each of which can be displayed in one of eight possible positions. The two flags cannot occupy the same position. How many symbols can be encoded in this way, remembering that the coding function must be an injection?

1.5 Coding for reliability

It is frequently necessary to send messages through unreliable channels, and in such circumstances we should like to use a method of coding that will reduce the likelihood of a mistake. An obvious technique is simply to repeat the message.

For example, suppose an investor communicates with a broker by sending the symbols B and S ($B = Buy$ and $S = Sell$). With this code, if any symbol is received incorrectly, the broker will make a mistake, and perform the wrong action.

However, suppose the investor uses the code $Buy \mapsto BB$ and $Sell \mapsto SS$. Now if any one symbol is received incorrectly the broker will know that something is wrong, because BS and SB are not codewords, and will be able to ask for further instructions.

If the investor uses more repetitions the broker may be able to make a reasonable decision about the intention, even when it is not possible to ask for further instructions. Suppose the investor uses the codewords BBB and SSS. Then, if SSB is received, it is more likely that the message was SSS, because that would imply that only one error had occurred, whereas BBB would imply that two errors had occurred.

In Chapters 6-9 we shall describe more efficient methods of coding messages so that the probability of a mistake due to errors in transmission is reduced.

EXERCISES

1.14. Suppose an investor uses the 5-fold repetition code, that is, $Buy \mapsto BBBBB$, $Sell \mapsto SSSSS$. If the following messages are received, which instruction is more likely to have been sent in each case?

$$BBBSB \qquad SBSBS \qquad SSSSB$$

1.15. Suppose we wish to send the numbers $1, 2, 3, 4, 5, 6$, representing the outcomes of a throw of a die, using binary codewords, all of the same length. What is the smallest possible length of the codewords? Suppose it is required that the receiver will notice whenever one bit in any codeword is in error. Find a set of codewords with length four which has this property.

1.6 Coding for security

One of the oldest codes is said to have been used by Julius Caesar over two thousand years ago, with the intention of communicating secretly with his army commanders. For a message in the 27-symbol alphabet $\mathbb{A}$, the rule is:

> *choose a number k between 1 and 25 and replace each letter by the one that is k places later, in alphabetical order.*

The rule is extended in an obvious way to the letters at the end of the alphabet, as in the example below. The space ⊔ is not changed. Thus if $k = 5$ the symbols are replaced according to the rule:

A B C D E F G H I J K L M N O P Q R S T U V W X Y Z ⊔
F G H I J K L M N O P Q R S T U V W X Y Z A B C D E ⊔ .

For example, the message

SEE⊔YOU⊔TOMORROW becomes XJJ⊔DTZ⊔YTRTWWTB .

In mathematical terms the coding rule is a function $c_k : \mathbb{A} \to \mathbb{A}$, which depends on the *key* k: in the example given above, $k = 5$. It is a basic assumption of cryptography that, although the value of k may be kept secret, the general form of the coding rule cannot. In other words, it will become known that the rule is c_k (apply a shift of k to the letters) for some k.

When a coded message such as

XJJ⊔DTZ⊔YTRTWWTB

is sent, it is presumed that the intended recipient knows the key – the value $k = 5$ in our example. In that case it is easy to decode the message. On the other hand, if someone who does not know the key intercepts the message, decoding is not necessarily so easy. In cryptography, decoding by finding the value of the key k (or otherwise) is said to be *breaking* the system, and any method which may achieve this is an *attack*.

In fact, Caesar's system is not very secure, because there is a simple attack by the method known as *exhaustive search*. The only possible values of k are

$1, 2, 3, \ldots, 25$, and it is easy to try each of them in turn, until a meaningful message is found.

Example 1.12

Suppose we have intercepted the message

SGZNY␣OY␣MUUJ␣LUX␣EUA .

We suspect that Caesar's system is being used. How do we find the key?

Solution Trying the possible keys, beginning with $k = 1$ and $k = 2$ produces the following possibilities. Remember that if the key is k, we must go *back* k places to find the original message.

$k = 1$: RFYMX␣NX␣ ···
$k = 2$: QEXLW␣MW␣ ···

Thus the key is not 1 or 2, because if it were, the original message would not make sense. There is no need to 'decode' the whole message in order to establish this fact. So we must continue to work through the keys $k = 3, 4, \ldots, 25$, until a meaningful message is found.

EXERCISES

1.16. Find the original message in Example 1.12.

1.17. Could the following message have been sent by Julius Caesar himself?

ZLJB␣LK␣BKDIXKA

1.18. Caesar's system is an example of a *substitution* code, because each letter in the message is replaced by a substitute letter, according to a fixed rule. Suggest other substitution rules, with a view to defending against the attack by exhaustive search.

Further reading for Chapter 1

The internet is a treasury of information about Morse code, semaphore, and other historically important coding systems. The pioneering work of Claude Shannon on the theory of information and communication is also well-represented.

Internet sites relating to cryptography are very variable in quality, and it is better to rely on good books such as those by Kahn [**1.2**] and Singh [**1.3**]. Older books on cryptography can also provide an important perspective for understanding the modern approach. The books by d'Agapeyeff [**1.1**] and Sacco [**1.4**] are recommended.

Books about the so-called 'Bible Codes' and similar matters should be regarded as entertainment. They are more entertaining (often unintentionally) when considered from the viewpoint of an informed reader, such as someone who has studied this book.

1.1 A. d'Agapeyeff. *Codes and Ciphers.* Oxford University Press, London (1939).

1.2 D. Kahn. *The Codebreakers.* Scribner, New York (1996).

1.3 S. Singh. *The Code Book.* Fourth Estate, London (2000).

1.4 L. Sacco. *Manuel de Cryptographie.* Payot, Paris (1951).

2
Prefix-free codes

2.1 The decoding problem

The symbols in a message appear in a certain order. So we often think of a message as part of a *stream*

$$\xi_1\xi_2\xi_3\cdots \quad ,$$

with the order being determined by a process that takes place in real time. Here each ξ_k is a variable that can take as its value any symbol from the alphabet S, and its actual value is the symbol that occurs at time k ($k = 1, 2, 3, \ldots$).

In this chapter we study the basic facts about coding and decoding a stream of symbols. We shall say that a coding function $c : S \to T^*$ replaces the *original stream* of symbols belonging to the alphabet S by an *encoded stream* of symbols belonging to the alphabet T.

Example 2.1

Let $S = \{x, y, z\}$ and suppose the original stream is

$$yzxxzyxzyy \cdots \quad .$$

The code $c : S \to \mathbb{B}^*$ is defined by the rules

$$x \mapsto 0, \qquad y \mapsto 10, \qquad z \mapsto 11.$$

What is the encoded stream?

N. L. Biggs, *An Introduction to Information Communication and Cryptography*,
DOI: 10.1007/978-1-84800-273-9_2,

Solution The encoded stream is simply the concatenation of the codewords for the individual symbols:

$$10110011100111010 \cdots \quad .$$

It is reasonable to require that the code c is uniquely decodable (Definition 1.11), so that the encoded stream corresponds to a unique original stream. Better still, we should like to decode the encoded stream in a sequential fashion, without having to wait for the message to be complete before we begin. This corresponds (very roughly) to what we do when we read a written message: we decode word-by-word and sentence-by-sentence. In that situation the decoding process is made easier by the presence of spaces (indicating end-of-word) and full stops (indicating end-of-sentence). It must be stressed that, in general, such helpful indications will not be available.

Here is an example in which sequential decoding is possible.

Example 2.2

Suppose we receive the encoded stream

$$100011100 \cdots \quad ,$$

and we are provided with a 'codebook' that tells us that the coding rule is as in Example 2.1. That is,

$$x \mapsto 0, \qquad y \mapsto 10, \qquad z \mapsto 11.$$

How do we find the original stream?

Solution The first symbol 1 is not in the codebook, so we look at the next symbol, 0. The string 10 is in the codebook as the code for y, so we decide that the first symbol of the original stream is y. The next symbol 0 is the codeword for x, so we decide that the next symbol is x. Continuing in this way we obtain the original stream $yxxzy \cdots$.

In general, for a code $c : S \to T^*$ the method used in Example 2.2 is as follows. Examine the symbols in the encoded stream in order, until a string q that is a complete codeword is recognized. Since c is an injection (by definition), there is a unique symbol $s \in S$ such that $c(s) = q$. So we decode this part of the stream as s, delete it, and repeat the process.

Clearly, this method will fail if the codeword q is also the *prefix* of another codeword q': that is, if $q' = qr$ for some nonempty word $r \in T^*$. In that case, we cannot distinguish between the complete word q and the initial part of q'.

Definition 2.3 (Prefix-free)

We say that a code $c : S \to T^*$ is *prefix-free* (or simply *PF*) if there is no pair of codewords $q = c(s)$, $q' = c(s')$ such that

$$q' = qr, \quad \text{for some nonempty word } r \in T^*.$$

Example 2.4

Let $S = \{w, x, y, z\}$ and define $c : S \to \mathbb{B}^*$ by

$$w \mapsto 10, \quad x \mapsto 01, \quad y \mapsto 11, \quad z \mapsto 011.$$

Is this code PF, and is it uniquely decodable?

Solution It is clearly not PF, since $c(x) = 01$ is a prefix of $c(z) = 011$. In fact it is not uniquely decodable either. For example, the string 10011011 corresponds to two different messages, $wxwy$ and wzz:

$$wxwy \mapsto 10\ 01\ 10\ 11 \qquad wzz \mapsto 10\ 011\ 011.$$

(The extra spaces are shown only to clarify the explanation.)

This example shows that a code that is not PF need not be UD. However, it is almost obvious that any PF code is UD. The formal proof is based on the sequential decoding method used in Example 2.2.

Theorem 2.5

If a code $c : S \to T^*$ is prefix-free, then it is uniquely decodable.

Proof

Suppose that $x_1x_2 \cdots x_m$ and $y_1y_2 \cdots y_n$ are strings in S^* such that their codes $c(x_1x_2 \cdots x_m)$ and $c(y_1y_2 \cdots y_n)$ are the same. That is

$$c(x_1)c(x_2) \cdots c(x_m) = c(y_1)c(y_2) \cdots c(y_n).$$

Since these strings are the same, their initial parts are the same. It follows that if $c(x_1) \neq c(y_1)$ then one is a prefix of the other, contrary to hypothesis. Hence $c(x_1) = c(y_1)$, and since c is an injection, $x_1 = y_1$.

Thus the remaining parts of the two strings are the same. Repeating the same argument, it follows that $x_2 = y_2$, and so on. Hence $x_1x_2 \cdots x_m = y_1y_2 \cdots y_n$ (and $m = n$), so c is UD. □

At first sight, it would seem that PF codes are very special. There are many possible decoding rules, and a code that is UD need not be PF. (See Exercise 2.3.) However, there are good reasons why PF codes are the only ones we need consider seriously. In the rest of this chapter we shall explain why this is so.

EXERCISES

2.1. A prefix-free binary code is defined by

$$s_1 \mapsto 00,\ s_2 \mapsto 010,\ s_3 \mapsto 100,\ s_4 \mapsto 111.$$

If the encoded stream is

$$10011101000111010 10000,$$

what is the original stream?

2.2. In Exercise 1.7 we found that the binary code

$$s_1 \mapsto 10, \quad s_2 \mapsto 010, \quad s_3 \mapsto 100$$

is not uniquely decodable. What is the underlying reason for this?

2.3. In the previous exercise, replace the coding of s_2 by $s_2 \mapsto 1$. Show that although the new code is not prefix-free, it is uniquely decodable.

2.4. Explain carefully why the Morse code, as defined in Section 1.4, is prefix-free. Explain also why the modified version (without $\odot$) discussed in Exercise 1.12 is not prefix-free.

2.2 Representing codes by trees

There is a useful way of representing a code by means of a *tree*. It works for codes in an alphabet of any size b, but for the purposes of illustration we shall focus on the binary alphabet $\mathbb{B}$ with $b = 2$.

First we observe that the set of nodes of an infinite binary tree (see Figure 2.1) can be regarded as $\mathbb{B}^*$, the set of all words in $\mathbb{B}$. Precisely, the *root* of the tree is labelled by the empty word, and the other nodes are labelled recursively, so that the two 'children' of the node labelled w are labelled $w0$ and $w1$.

We shall follow the standard (but perverse) practice of using a mixture of botanical and genealogical terms to describe a tree. For example, these trees 'grow' downwards, with the root at the top 'level'.

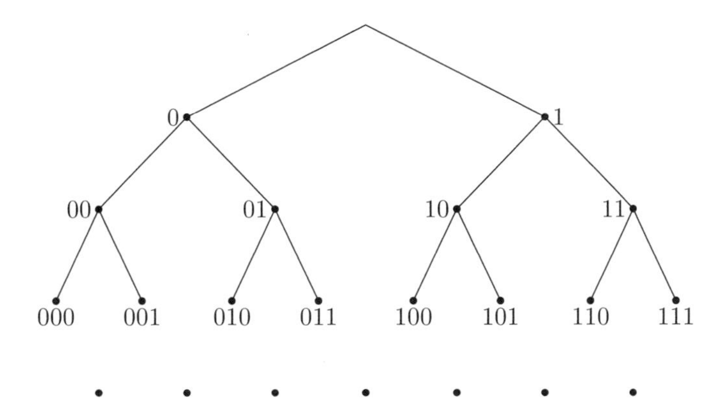

Figure 2.1 The nodes of the infinite binary tree correspond to $\mathbb{B}^*$

Now suppose that a code $C \subseteq \mathbb{B}^*$ is given, and the codewords are represented by the corresponding nodes of the binary tree. Clearly, a codeword q is a prefix of another codeword q' if the unique path from q' to the root of the tree passes through q.

If the code C is prefix-free it follows that, for any codeword q, none of the descendants of q can be a codeword. Thus we can ignore all the nodes that are descendants of q. If we also ignore all those nodes that are neither codewords nor prefixes of codewords, we have a finite binary tree, and the codewords of C are its *leaves*.

Example 2.6

Represent the code $C = \{0, 10, 110, 111\}$ by means of a tree.

Solution The tree is shown in Figure 2.2.

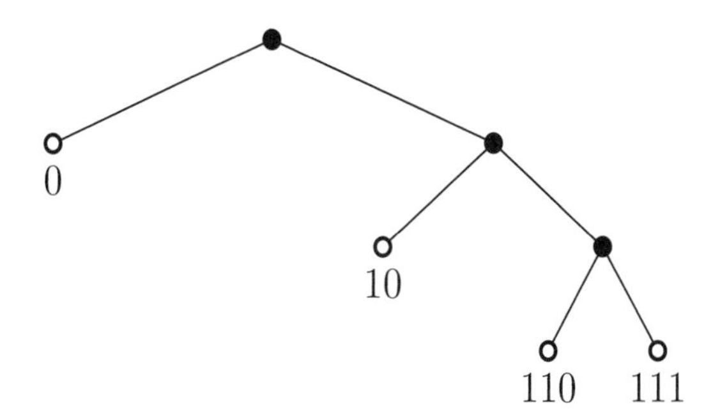

Figure 2.2 A set of codewords represented as the leaves on a tree

EXERCISES

2.5. Sketch the tree that represents the PF code

$$s_1 \mapsto 00,\ s_2 \mapsto 010,\ s_3 \mapsto 100,\ s_4 \mapsto 111,$$

and label each leaf with the corresponding symbol. Is it possible to extend this code without destroying the PF property?

2.6. Denote the *ternary* alphabet by $T = \{0, 1, 2\}$. Construct a tree representing the ternary code with codewords $02, 101, 120, 221, 222$. Is it possible to extend this code without destroying the PF property?

2.7. What is the maximum number of codewords in a prefix-free binary code in which the longest codeword has length 7?

2.8. In Exercise 1.15 we discussed the problem of representing the numbers $1, 2, 3, 4, 5, 6$ by a binary code, in such a way all six codewords have the same length. Suppose we now ask for a PF binary code, such that the total length of the codewords is as small as possible. Construct a suitable code, using the tree representation as a guide.

2.3 The Kraft-McMillan number

Given a code $c : S \to T^*$, let n_i denote the number of symbols in S that are encoded by strings of length i in T^*. In other words, n_i is the number of $s \in S$ such that $c(s)$ is in T^i. If M is the maximum length of a codeword, we refer to the numbers $n_1, n_2, \ldots, n_M$ as the *parameters* of c. In Figure 2.2 the parameters are $n_1 = 1$, $n_2 = 1$, $n_3 = 2$. In general, in the tree diagram, n_i is the number of codewords at level i.

When $|T| = b$ the total number of strings of length i in T^* is $|T^i| = b^i$. By definition, a code $c : S \to T^*$ is an injection, so its parameters must satisfy $n_i \leq b^i$. The fraction n_i/b^i represents the proportion of words of length i that are used as codewords.

Definition 2.7 (Kraft-McMillan number)

The *Kraft-McMillan number* associated with the parameters $n_1, n_2, \ldots, n_M$, and the base b, is defined to be

$$K = \sum_{i=1}^{M} \frac{n_i}{b^i} = \frac{n_1}{b} + \frac{n_2}{b^2} + \cdots + \frac{n_M}{b^M}.$$

For example, if $n_1 = 1$, $n_2 = 2$, $n_3 = 1$, and $b = 2$, then

$$K = \frac{1}{2} + \frac{2}{4} + \frac{1}{8} = \frac{9}{8}.$$

We shall prove two important results about the existence of b-ary codes.

- If $K \leq 1$ for a set of parameters $n_1, n_2, \ldots, n_M$ then there is a PF b-ary code with those parameters.
- The parameters of a UD b-ary code must satisfy the condition $K \leq 1$.

Taken together, these two results imply that if there is a UD code with certain parameters, then there is PF code with the same parameters.

The proof of the first result is best explained in terms of the tree representation. The basic idea is illustrated in the following example.

Example 2.8

Construct a PF binary code with parameters $n_2 = 2$, $n_3 = 3$, $n_4 = 1$.

Solution First, we check that the $K \leq 1$ condition is satisfied:

$$K = \frac{2}{4} + \frac{3}{8} + \frac{1}{16} = \frac{15}{16} \leq 1.$$

For the construction, since $n_2 = 2$ we start with two codewords of length 2, say 00 and 01. The PF condition means that we cannot use any words of length 3 the form $0\,0*$ or $0\,1*$, but there remain four other possible words, of the form $1**$. In fact we require only three of them, so we can choose 100, 101, 110, for example. Finally, we require one word of length 4, which can be either 1110 or 1111.

Theorem 2.9

If $K \leq 1$ for the parameters $n_1, n_2, \ldots, n_M$, then a PF b-ary code with these parameters exists.

Proof

First, note that in the inequality

$$K = \frac{n_1}{b} + \frac{n_2}{b^2} + \cdots + \frac{n_M}{b^M} \leq 1,$$

each term n_i/b^i is non-negative. This means that any partial sum of these terms is also not greater than 1:

$$\frac{n_1}{b} \leq 1, \quad \frac{n_1}{b} + \frac{n_2}{b^2} \leq 1, \quad \frac{n_1}{b} + \frac{n_2}{b^2} + \frac{n_3}{b^3} \leq 1, \quad \cdots .$$

These inequalities can be rewritten as

$$n_1 \leq b, \quad n_2 \leq b(b - n_1), \quad n_3 \leq b(b^2 - n_1 b - n_2), \quad \cdots ,$$

and generally,

$$n_i \leq b(b^{i-1} - n_1 b^{i-2} - \cdots - n_{i-1}).$$

Since $n_1 \leq b$, it is possible to choose n_1 different codewords of length 1. There remain $b - n_1$ words of length 1, each of which gives rise to b words of length 2, and the PF condition means that only these $b(b-n_1)$ words of length 2 are available as codewords. Since $n_2 \leq b(b - n_1)$, it is possible to choose n_2 of them as codewords (Figure 2.3).

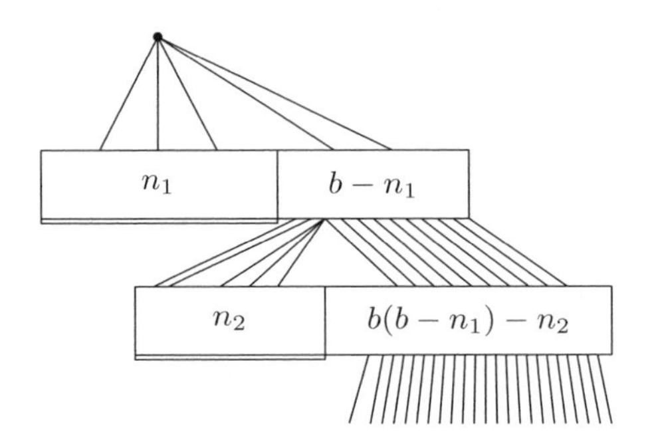

Figure 2.3 Finding codewords when $K \leq 1$: initial steps

As before, the PF condition means that only $b(b^2 - n_1 b - n_2)$ words of length 3 are available as codewords. Since $n_3 \leq b(b^2 - n_1 b - n_2)$, it is possible to choose n_3 of them.

The general step is illustrated in Figure 2.4. The number of unused codewords of length $i - 1$ is

$$f_{i-1} = b^{i-1} - n_1 b^{i-2} - \cdots - n_{i-1}$$

and since $n_i \leq b f_{i-1}$, it is possible to choose n_i words of length i as codewords without violating the PF condition. Thus the condition $K \leq 1$ ensures that enough codewords are available at each step, and a PF code with the given parameters can be constructed. □

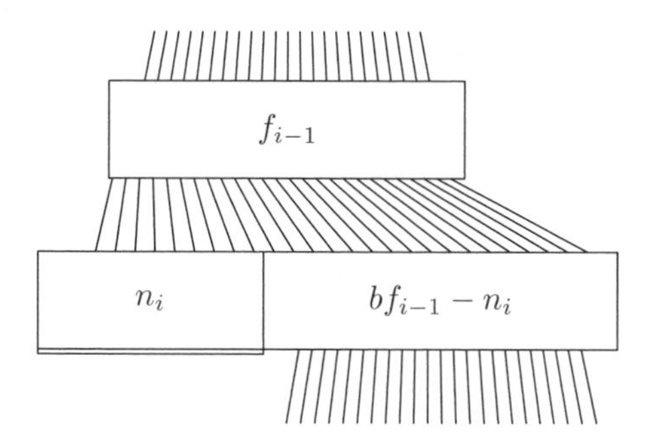

Figure 2.4 Finding codewords when $K \leq 1$: general step

EXERCISES

2.9. Construct PF binary codes with the following parameters:

(i) $n_2 = 1$, $n_3 = 4$, $n_4 = 3$;

(ii) $n_1 = 1$, $n_2 = 0$, $n_3 = 2$, $n_4 = 3$, $n_5 = 2$.

2.10. On the basis of Theorem 2.9, what can be said about the existence of prefix-free ternary ($b = 3$) codes with the following parameters?

(i) $n_1 = 0$, $n_2 = 1$, $n_3 = 12$;

(ii) $n_1 = 0$, $n_2 = 1$, $n_3 = 12$, $n_4 = 40$.

2.11. If the conclusion of the previous exercise is that a prefix-free code must exist, construct one.

2.12. Let us say that a code is *complete* if it is PF and $K = 1$. Show that if there is a complete b-ary code for an alphabet of size m, then $b-1$ must divide $m - 1$.

2.4 Unique decodability implies $K \leq 1$

In this section we prove that the parameters of a UD code must satisfy $K \leq 1$. The proof involves some elementary combinatorial theory.

Let $c : S \to T^*$ be a code with parameters $n_1, n_2, \ldots, n_M$. Since M is the length of the longest codeword for a single symbol, the maximum length of the code for a string of r symbols is rM.

Given $r \geq 1$, let $q_r(i)$ be the number of strings of length r that are encoded by strings of length i $(1 \leq i \leq rM)$; in particular, $q_1(i) = n_i$. We shall show that the parameters n_i determine the numbers $q_r(i)$, for all r.

Example 2.10

Suppose $c : S \to T^*$ is a code with parameters $n_1 = 2$, $n_2 = 3$. What are the numbers $q_2(i)$?

Solution Let xy be a string of length 2 in S. Since $c(x)$ and $c(y)$ have length 1 or 2, $c(xy)$ has length $2, 3$, or 4.

If $c(xy)$ has length 2, $c(x)$ and $c(y)$ must both have length 1. Since $n_1 = 2$, there are 2×2 possible strings xy. Thus $q_2(2) = 4$.

If $c(xy)$ has length 3, $c(x)$ must have length 1 or 2, and $c(y)$ must have length 2 or 1, respectively. Since $n_1 = 2$ and $n_2 = 3$, there are $2 \times 3 + 3 \times 2$ possible strings xy. Thus $q_2(3) = 6 + 6 = 12$.

If $c(xy)$ has length 4, $c(x)$ and $c(y)$ must both have length 2. Since $n_2 = 3$, there there are 3×3 possible strings xy. Thus $q_2(4) = 9$.

We define the *generating function* for the sequence $q_r(1), q_r(2), \ldots q_r(rM)$, as follows:

$$Q_r(x) = q_r(1)x + q_r(2)x^2 + \cdots + q_r(rM)x^{rM}.$$

Example 2.11

What is the relationship between $Q_1(x)$ and $Q_2(x)$ for the code described in the previous example?

Solution The generating functions for $r = 1$ and $r = 2$ are

$$Q_1(x) = 2x + 3x^2, \qquad Q_2(x) = 4x^2 + 12x^3 + 9x^4,$$

so that $Q_2(x) = Q_1(x)^2$.

This example suggests the following result, which we refer to as the *Counting Principle.* The proof will be given in Section 2.5.

Lemma 2.12

For all $r \geq 1$, $Q_r(x) = Q_1(x)^r$.

Assuming this result, we can prove the second important theorem about the Kraft-McMillan number.

Theorem 2.13

If a uniquely decodable b-ary code exists, with a given set of parameters, then its Kraft-McMillan number satisfies $K \leq 1$.

Proof

Fix $r \geq 1$. Putting $x = 1/b$ in the definition of $Q_r(x)$ we have

$$Q_r(1/b) = q_r(1)/b + q_r(2)/b^2 + \cdots + q_r(rM)/b^{rM}.$$

Unique decodability implies that $q_r(i)$ cannot exceed b^i, the total number of strings of length i. It follows that every term in $Q_r(1/b)$ is less than or equal to 1, and since the number of terms is at most rm, $Q_r(1/b) < rM$.

Now the Kraft-McMillan number is given by

$$K = \frac{n_1}{b} + \frac{n_2}{b^2} + \cdots + \frac{n_M}{b^M} = Q_1(1/b).$$

Thus, by the Counting Principle

$$K^r = Q_1(1/b)^r = Q_r(1/b).$$

Since $Q_r(1/b) < rM$ it follows that $K^r/r < M$ for all r, where M is a fixed constant.

Suppose it were true that $K > 1$, so that $K = 1 + h$ with $h > 0$. By the binomial theorem,

$$K^r = (1+h)^r = 1 + rh + \frac{1}{2}r(r-1)h^2 + \cdots .$$

Since all terms are positive, K^r is certainly greater than the term in h^2, that is

$$\frac{K^r}{r} > \frac{1}{2}(r-1)h^2.$$

Thus by taking r sufficiently large, we could make K^r/r as large as we please, contrary to the result obtained above. So we must have $K \leq 1$. □

Theorem 2.13 can be combined with Theorem 2.9 as follows:

$$\text{UD code exists} \implies K \leq 1 \implies \text{PF code exists.}$$

The combined result is that the existence of a UD code with given parameters implies the existence of a PF code with the same parameters. We can therefore say that 'PF codes suffice'. The converse of this result was proved in Theorem 2.5. So we conclude that there is a UD code with certain parameters *if and only if* there is a PF code with the same parameters, and this happens *if and only if* $K \leq 1$.

EXERCISES

2.13. Suppose we wish to construct a UD code for 12 symbols using binary words of length not exceeding 4. Make a list of all the sets of parameters n_1, n_2, n_3, n_4 for which a suitable code exists.

2.14. Let $S = \{s_1, s_2, \ldots, s_6\}$ and $T = \{a, b, c\}$, and suppose that a code is defined by

$$s_1 \mapsto a, \ s_2 \mapsto ba, \ s_3 \mapsto bb, \ s_4 \mapsto bc, \ s_5 \mapsto ca, \ s_6 \mapsto cb.$$

Write down the generating function $Q_1(x)$ and hence (by algebraic means) find $Q_2(x)$.

2.15. In the previous exercise, what does the coefficient of x^4 in $Q_2(x)$ represent? Verify your answer by making a list of the corresponding elements of S^*.

2.16. Let $S = \{a, b, c, d, e, f, g\}$ and suppose that a binary code is defined by

$$a \mapsto 00, \ b \mapsto 010, \ c \mapsto 011, \ d \mapsto 1000,$$

$$e \mapsto 1001, \ f \mapsto 1101, \ g \mapsto 1111.$$

Write down the generating function $Q_1(x)$ and hence calculate $Q_2(x)$ and $Q_3(x)$. What do the coefficients of x^7 in $Q_2(x)$ and $Q_3(x)$ represent? Verify your answers by making lists of the corresponding elements of S^*.

2.5 Proof of the Counting Principle

Theorem 2.14 (The Counting Principle)

Let $c : S \to T^*$ be a code such that, for all $s \in S$, the length of $c(s)$ is not greater than M. Given $r \geq 1$, let $q_r(i)$ be the number of strings of length r in S that are encoded by strings of length i in T ($1 \leq i \leq rM$). Let $Q_r(x)$ be the generating function

$$Q_r(x) = q_r(1)x + q_r(2)x^2 + \cdots + q_r(rM)x^{rM}.$$

Then $Q_r(x) = Q_1(x)^r$.

Proof

We consider the coefficient of x^N in the product

$$Q_1(x)^r = (n_1 x + n_2 x^2 + \cdots + n_M x^M)^r,$$

where, as usual, $n_i = q_1(i)$.

A contribution to the coefficient of x^N is obtained by multiplying r terms, one from each of the r factors. That is, we choose a term $n_{i_1} x^{i_1}$ from the first factor, $n_{i_2} x^{i_2}$ from the second factor, and so on, in such a way that $i_1 + i_2 + \cdots + i_r = N$, so that the resulting power of x is N. Hence the coefficient of x^N is the sum of all products

$$n_{i_1} n_{i_2} \ldots n_{i_r} \qquad \text{where} \quad i_1 + i_2 + \cdots + i_r = N.$$

Now consider $q_r(N)$, the number of strings of length N that can be formed from the strings representing r symbols in S. Such a string is formed by choosing any one of n_{i_1} codewords of length i_1, n_{i_2} codewords of length i_2, and so on, in such a way that $i_1 + i_2 + \cdots + i_r = N$. Hence $q_r(N)$ is just the coefficient of x^N, as in the previous paragraph. □

EXERCISES

2.17. For a given code $c : S \to T^*$, let $C(x, y)$ be the two-variable generating function for the numbers $q_r(i)$ defined above, that is

$$C(x, y) = \sum_{i,r} q_r(i) x^i y^r.$$

Prove that $C(x, y) = yQ_1(x)/(1 - yQ_1(x))$.

Further reading for Chapter 2

The number that we have denoted by K, and called the 'Kraft-McMillan' number, has an interesting history. Our Theorem 2.9 ($K \leq 1$ implies that a PF code exists) was proved by L.G. Kraft in his 1949 Master's thesis [**2.3**]. He also gave a proof of the converse result. The converse also follows from our Theorem 2.13 (UD implies that $K \leq 1$), which was first proved by B. McMillan in 1956 [**2.4**] using a method based on complex variable theory. A simpler proof was discovered by J. Karush in 1961 [**2.2**], and our proof is based on Karush's method.

The fact that it is an application of a standard result on generating functions is often overlooked.

A good account of the early history of this topic is available in the collection of articles edited by E.R. Berlekamp [**2.1**].

2.1 E.R. Berlekamp. *Key Papers in the Development of Coding Theory.* IEEE Press, New York (1974).

2.2 J. Karush. A simple proof of an inequality of McMillan. *IRE Trans. Information Theory* IT-7 (1961) 118.

2.3 L.G. Kraft. *A device for quantizing grouping, and coding amplitude modulated pulses*, M.S. thesis, Electrical Engineering Department, MIT (1949).

2.4 B. McMillan. Two inequalities implied by unique decipherability. *IRE Trans. Information Theory* IT-2 (1956) 115-116.

3
Economical coding

3.1 The concept of a source

Roughly speaking, a 'source' is a means of producing messages. Examples are a human being writing email messages, or a scanner making digitized images. Using the terminology introduced in Chapter 2, we say that a source *emits a stream* of symbols, denoted by

$$\xi_1 \xi_2 \xi_3 \cdots \quad .$$

Here each ξ_k is a variable that can take as its value any symbol belonging to a given alphabet.

An obvious feature that distinguishes one source from another is the alphabet. Writers in oriental languages use alphabets that are quite different from the 27-letter alphabet that we have denoted by $\mathbb{A}$. However, there are other distinctive features. Writers in western languages all use alphabets similar to $\mathbb{A}$, but in different ways. In a natural language there are certain rules about how the alphabet is used (spelling, grammar, syntax) and these are reflected in the stream emitted by the source. For example, in English the symbol `J` will occur much less often than the symbol `E`. Similarly, if a scanner is set up to look at text documents, it will produce messages in which W (representing a white pixel) is much more likely than B (representing a black pixel).

These remarks lead to the conclusion that a source can be described by specifying both the alphabet and the probabilities of the symbols in the stream emitted. Suppose, for example, that the source is an experiment in which a coin is tossed repeatedly, the results being recorded as $H = Head$, $T = Tail$.

N. L. Biggs, *An Introduction to Information Communication and Cryptography*,
DOI: 10.1007/978-1-84800-273-9_3,

If the coin is a fair one, the stream emitted might be

$$HTTHTHTHTTHHTTTHTHHTTHTHTHTH\ldots,$$

where H and T occur with roughly the same frequency. This is a source with alphabet $S = \{H, T\}$. The probability that ξ_k is H, and the probability that ξ_k is T, are both $\frac{1}{2}$. That is, for all k,

$$\Pr(\xi_k = H) = \Pr(\xi_k = T) = 0.5.$$

On the other hand, an unfair coin would emit a stream such as

$$HHHHHHHTHHHHHTHHHHHHHHHHT\ldots,$$

where the probabilities are such that $\Pr(\xi_k = H)$ is significantly greater than $\Pr(\xi_k = T)$.

The theory required to set up a complete mathematical model of a source is, in general, fairly complicated, but we can make some progress by looking at a very simple model.

Definition 3.1 (Probability distribution)

Let $S = \{s_1, s_2, \ldots, s_m\}$ be an alphabet. A *probability distribution* on S is a set of real numbers $p_1, p_2, \ldots, p_m$ such that

$$p_1 + p_2 + \ \ldots\ + p_m = 1, \qquad 0 \le p_i \le 1\ (i = 1, 2, \ldots, m).$$

We shall usually write the numbers p_i as a row vector $\mathbf{p} = [p_1, p_2, \ldots, p_m]$, and say that the symbol s_i occurs with probability p_i.

Given a probability distribution $\mathbf{p}$ on S, consider a source with the following property. The source emits a stream $\xi_1\xi_2\xi_3\cdots$, where the value of each ξ_k is a member of the alphabet S and, for all k, the probability that ξ_k takes a particular value s_i is given by

$$\Pr(\xi_k = s_i) \ = \ p_i.$$

Technically, ξ_k is a *random variable.*

We shall refer to this model as a 'source with probability distribution $\mathbf{p}$' or simply a 'source $(S, \mathbf{p})$'. It is important to stress that the terminology does not imply that all the characteristics of the source are described by $\mathbf{p}$, it simply means that the probability distribution for each individual ξ_k is the same.

In many cases this is a reasonable assumption, even though the source may have other, more subtle, features. For example, suppose we regard the text of a book as a stream of symbols in the alphabet $\mathbb{A}$. If we open the book at page x and find the yth symbol on that page, then the probability that the symbol is E can reasonably be assumed to be the same, for all x and y. But if we know that the 17th letter on page 83 is Q, then that affects the probability that the 18th letter on page 83 is E. In technical terms, we are imposing the condition that the random variables ξ_k all have the same distribution $\mathbf{p}$ but, in general, we do not assume that these random variables are *independent*.

Definition 3.2 (Memoryless source)

A source $(S, \mathbf{p})$ that emits a stream $\xi_1\xi_2\xi_3\cdots$ is *memoryless* if the random variables ξ_k are independent. That is, for all k and ℓ

$$\Pr(\xi_k = s_i \text{ and } \xi_\ell = s_j) \;=\; \Pr(\xi_k = s_i)\Pr(\xi_\ell = s_j).$$

For a memoryless source, the distribution $\mathbf{p}$ provides a complete description: knowledge of any of the terms in a message does not affect the probability distributions assigned to the other terms. This property can be assumed to hold in the coin-tossing experiment described above, and in some other real situations.

Example 3.3

In the UK there is a weekly ritual called the 'football results'. This can be regarded as a source that emits a stream of symbols chosen from the alphabet $S = \{h, a, d\}$, where h = home win, a = away win, d = draw. Is this a memoryless source?

Solution It is reasonable to assume that this source is memoryless: the result of one football match does not affect the result of any other matches in the list. Observation suggests that the probability distribution is, roughly,

$$p_h = 0.42, \quad p_a = 0.26, \quad p_d = 0.31.$$

Again, it must be stressed that many real sources are not memoryless. We shall return to this point in the next chapter. Even in the memoryless case, there are questions about how probabilities should be interpreted in real situations. The coin-tossing example indicates that it may be necessary to infer the characteristics of the source by examining the stream that it emits. The purpose of the experiment might be to check whether or not the coin is a fair one. We shall generally assume that the probability of an event represents the relative frequency of its occurrence, but for some purposes it is helpful to adopt an alternative viewpoint in which the probability is regarded as a 'degree of belief' (the *Bayesian approach*).

EXERCISES

3.1. Consider a source emitting symbols from the alphabet S with probability distribution $\mathbf{p}$, where

$$S = \{a, b, c\} \qquad \mathbf{p} = [p_a, p_b, p_c] = [0.6, 0.3, 0.1].$$

If this source emits a stream of 100 symbols, approximately how many times will a occur? If the source is memoryless, approximately how may times will the consecutive pair ab occur?

3.2. An instrument records the temperature in degrees Celsius every hour at a certain place. Is this a memoryless source?

3.3. A source emits the stream of positive integers $\xi_1\xi_2\xi_3\cdots$ defined by the rules $\xi_1 = 1$, $\xi_2 = 1$, $\xi_k = \xi_{k-1} + \xi_{k-2}$ $(k \geq 3)$. How does this source differ from the ones discussed in this section?

3.2 The optimization problem

We now consider what happens when the *original stream* emitted by a source $(S, \mathbf{p})$ is transformed into an *encoded stream*, using an alphabet T. In particular, how does the length of a typical message depend on the code $c : S \to T^*$ that is used?

Suppose that $S = \{s_1, s_2, \ldots, s_m\}$. Let y_i be the length of the codeword $c(s_i)$ $(1 \leq i \leq m)$, and consider a message of length N in S. If N is a reasonably large number, the message contains the symbol s_1 approximately Np_1 times, s_2 approximately Np_2 times, and so on. After encoding with c, there are Np_1 strings $c(s_1)$, each of length y_1, Np_2 strings $c(s_2)$, each of length y_2, and so on. The total length of the encoded message is approximately

$$Np_1y_1 + Np_2y_2 + \cdots Np_my_m \;=\; N(p_1y_1 + p_2y_2 + \;\cdots\; + p_my_m).$$

Thus if the original message has length N, the encoded message has length LN, approximately, where $L = p_1y_1 + p_2y_2 + \;\cdots\; + p_my_m$.

Definition 3.4 (Average word-length)

The *average word-length* of a code $c : S \to T^*$ for the source $(S, \mathbf{p})$ is

$$L = p_1y_1 + p_2y_2 + \;\cdots\; + p_my_m.$$

Example 3.5

Let $S = \{s_1, s_2, s_3\}$ and $\mathbf{p} = [0.2, 0.6, 0.2]$. Find the value of L when the binary code

$$s_1 \mapsto 0, \quad s_2 \mapsto 10, \quad s_3 \mapsto 11$$

is used. Is there a binary code for S that achieves a smaller value?

Solution The value of L is

$$1 \times 0.2 + 2 \times 0.6 + 2 \times 0.2 = 1.8.$$

Noting that s_2 is the most likely symbol, we should try a code that uses the shortest codeword for this symbol, such as

$$s_1 \mapsto 10, \quad s_2 \mapsto 0, \quad s_3 \mapsto 11.$$

Here the value of L is

$$2 \times 0.2 + 1 \times 0.6 + 2 \times 0.2 = 1.4.$$

In this example it is fairly easy to improve on the original code, but we did not ask whether further improvements are possible.

Question: Given the source $(S, \mathbf{p})$ and the alphabet T, how can we find a code $c : S \to T^*$ for which L is as small as possible?

For practical purposes, it is important that the code is uniquely decodable, so we make the following definition.

Definition 3.6 (Optimal code)

Given a source $(S, \mathbf{p})$ and an alphabet T, a UD code $c : S \to T^*$ is *optimal* if there is no such code with smaller average word-length.

The requirement that the code be UD is a significant constraint. In Chapter 2 we showed that this constraint can be expressed by the condition $K \leq 1$, where K is the Kraft-McMillan number associated with parameters $n_1, n_2, \ldots, n_M$ and the base b of the code. Explicitly

$$K = \frac{n_1}{b} + \frac{n_2}{b^2} + \cdots + \frac{n_M}{b^M}.$$

In our current notation, n_i is the number of symbols s_j such that $y_j = i$, and M is the maximum value of y_j. The term n_i/b^i in K is the sum of n_i terms

$1/b^i$, corresponding to one term $1/b^{y_j}$ for each j such that $y_j = i$. It follows that K can be written as the sum of all the terms $1/b^{y_j}$:

$$K = \frac{1}{b^{y_1}} + \frac{1}{b^{y_2}} + \cdots + \frac{1}{b^{y_m}}.$$

Rewriting the $K \leq 1$ condition in this form we can formulate the problem of finding optimal codes as follows.

Given b and $p_1, p_2, \ldots, p_m$ find positive integers $y_1, y_2, \ldots, y_m$ that

$$\text{minimize} \quad p_1 y_1 + p_2 y_2 + \cdots + p_m y_m$$

$$\text{subject to} \quad \frac{1}{b^{y_1}} + \frac{1}{b^{y_2}} + \cdots + \frac{1}{b^{y_m}} \leq 1.$$

This problem cannot be solved completely by 'calculus methods', due to the condition that the y_i's must be integers. In the following sections we shall discuss it using an approach based on one of the most important concepts in coding theory.

EXERCISES

3.4. A source emits three symbols with probabilities 0.5, 0.25, 0.25. Construct a PF binary code for this source with average word-length 1.5.

3.5. Consider the general case of a source emitting three symbols with probability distribution

$$[\alpha, \beta, 1 - \alpha - \beta] \quad \text{where} \quad \alpha > \beta > 1 - \alpha - \beta \geq 0.$$

Show that the average word-length of an optimal binary code for this source is $2 - \alpha$.

3.3 Entropy

Definition 3.7 (Entropy of a distribution)

The *entropy* to base b of a probability distribution $\mathbf{p} = [p_1, p_2, \ldots, p_m]$ is

$$H_b(\mathbf{p}) \;=\; H_b(p_1, p_2, \ldots, p_m) \;=\; \sum_{i=1}^{m} p_i \log_b(1/p_i).$$

If $0 < p_i < 1$ then $1/p_i > 1$, and each term $p_i \log_b(1/p_i)$ is positive. Occasionally it is convenient to allow $p_i = 0$, when the expression $p_i \log_b(1/p_i)$ is strictly not defined. However, we shall give it the conventional value 0, since that is its limit as $p_i \to 0$.

Since the output of a *memoryless* source $(S, \mathbf{p})$ is completely determined by $\mathbf{p}$, we shall often speak of the entropy of $\mathbf{p}$ as the entropy of the source. However, it must be stressed that many sources are not memoryless, and for them a more sophisticated definition of entropy is required (see Chapter 4).

Example 3.8

What is the entropy to base 2 of a memoryless source with distribution $\mathbf{p} = [0.5, 0.25, 0.25]$?

Solution

$$\begin{aligned} H_2(p_1, p_2, p_3) &= p_1 \log_2(1/p_1) + p_2 \log_2(1/p_2) + p_3 \log_2(1/p_3) \\ &= 0.5 \log_2(1/0.5) + 0.25 \log_2(1/0.25) + 0.25 \log_2(1/0.25) \\ &= 0.5 \log_2 2 + 0.25 \log_2 4 + 0.25 \log_2 4 \\ &= 0.5 \times 1 + 0.25 \times 2 + 0.25 \times 2 \\ &= 1.5. \end{aligned}$$

EXERCISES

3.6. What is the entropy to base 2 of a memoryless source emitting five letters `A,E,I,O,U` with probabilities 0.2, 0.3, 0.2, 0.2, 0.1? What is the entropy if all five letters are equally probable?

3.7. What is the entropy of a source that is certain to emit one specific symbol?

3.8. What is the entropy of a memoryless source that emits symbols from an alphabet of size m, each symbol being equally probable?

3.9. Suppose that m men and w women enter a 'reality TV' contest. The probability that the ith man will win is u_i and the probability that the jth woman will win is v_j. Let

$$U = u_1 + u_2 + \ldots u_m, \qquad V = v_1 + v_2 + \ldots + v_w.$$

Show that the distribution $[u_1, u_2, \ldots, u_m, v_1, v_2, \ldots, v_w]$ has entropy equal to

$$H(U,V) + UH(u_1/U, u_2/U, \ldots, u_m/U) + VH(v_1/V, v_2/V, \ldots v_w/V).$$

(This result has a simple interpretation: see Exercise 3.12.)

3.4 Entropy, uncertainty, and information

At this point the definition of entropy is unmotivated. What is its significance? What is the role of the base b? How is entropy relevant to the optimal coding problem?

In order to answer these questions, consider first a very simple example, a memoryless source emitting the two symbols 0 and 1. The probabilities p_0 and p_1 can be written as x and $1-x$ for some x in the range $0 \le x \le 1$, and the entropy of this source (to base 2) is

$$h(x) = x \log_2(1/x) + (1-x) \log_2(1/(1-x)).$$

This function turns up frequently, and we shall reserve the letter h for it. The graph of h for $0 \le x \le 1$ is shown in Figure 3.1. It is symmetrical about the line $x = \frac{1}{2}$ and the value $h(\frac{1}{2}) = 1$ is the maximum value (Exercise 3.11).

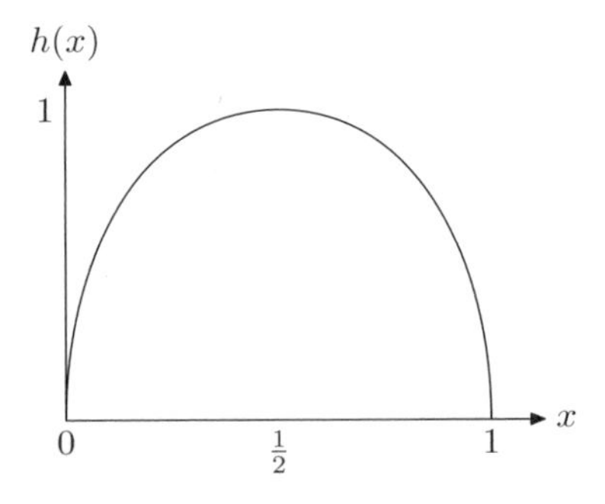

Figure 3.1 The graph of $h(x)$

Figure 3.1 suggests that entropy is a measure of *uncertainty*. We would expect that the greatest uncertainty about which symbol is emitted will occcur when the two symbols are equally probable ($x = \frac{1}{2}$); on the other hand there is no uncertainty if one of the symbols is never emitted ($x = 0$ or $x = 1$). In fact, it can be shown quite generally that the entropy

$$H_b(\mathbf{p}) = \sum p_i \log_b(1/p_i),$$

is a measure of the uncertainty about the identity of a symbol that is taken from a set according to the probability distribution $\mathbf{p}$. To be precise, $H_b(\mathbf{p})$ is essentially the only function that satisfies some very reasonable conditions that we associate with the notion of uncertainty.

The word 'essentially' is needed here because we can always change the unit of measurement, which corresponds to multiplying by a constant. In fact, this is already part of our definition, since the base b is not specified. If we chose another base a then the identity

$$\log_a x = \log_a b \times \log_b x$$

implies that

$$H_a(\mathbf{p}) = \log_a b \times H_b(\mathbf{p}).$$

Thus changing the base amounts to changing the unit of measurement. The base $b = 2$ is usually taken as the standard, in which case we write $H(\mathbf{p})$ instead of $H_2(\mathbf{p})$, and we say that $H(\mathbf{p})$ measures the uncertainty in *bits per symbol*. If it is appropriate to use a different base (as in the next Section) we can use the relation

$$H_b(\mathbf{p}) = \frac{H(\mathbf{p})}{\log_2 b}.$$

The concept of *information* is closely related to *uncertainty*. Roughly speaking, providing information about an event reduces our uncertainty about it. We can reconcile this idea with our definition of the entropy-uncertainty of $\mathbf{p}$ in the following way.

Suppose a memoryless source emits a steam of symbols $\xi_1\xi_2\xi_3\cdots$, where each ξ_k has the distribution $\mathbf{p}$. The entropy is the sum of terms $p_i \log(1/p_i)$, each of which is the product of the probability p_i that ξ_k has the value s_i and the quantity $\log(1/p_i)$. If we interpret $\log(1/p_i)$ as the amount of information provided by knowing that $\xi_k = s_i$, then the sum $\sum p_i \log(1/p_i)$ is, on average, the amount of information provided by knowing the value of ξ_k.

For example, when the source emits two symbols, with probabilities x and $1 - x$, the information provided is $h(x)$ bits per symbol. As we noted above, this is greatest when the symbols are equally probable. It is zero when one of the symbols has probability 0, because then we know the value of each ξ_k in advance.

We can now develop some of the mathematical properties of entropy-uncertainty-information. A very useful result (Theorem 3.10) relies on the following lemma from elementary calculus.

Lemma 3.9

For all $x > 0$, $\ln x \leq x - 1$, with equality if and only if $x = 1$. (Figure 3.2.)

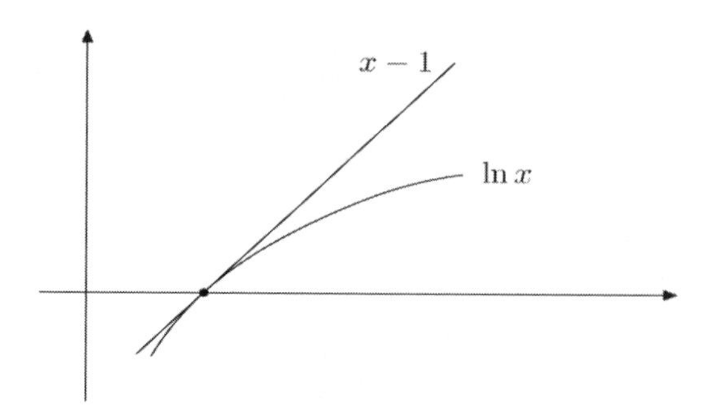

Figure 3.2 Graphs of $x - 1$ and $\ln x$

Proof

Let $f(x) = x - 1 - \ln x$. Then $f'(x) = 1 - 1/x$, which is zero only when $x = 1$. Since $f''(1) > 0$, the value $f(1) = 0$ is the minimum value of f, and the result follows. □

Theorem 3.10 (The comparison theorem)

If $p_1, p_2, \ldots, p_m$ and $q_1, q_2, \ldots, q_m$ are probability distributions then

$$H_b(\mathbf{p}) = \sum_{i=1}^{m} p_i \ \log_b(1/p_i) \leq \sum_{i=1}^{m} p_i \ \log_b(1/q_i).$$

There is equality if and only if $q_i = p_i$ for all i $(1 \leq i \leq m)$.

Proof

It is sufficient to prove the result for any one value of b, since changing b to b' amounts to multiplying both sides of the inequality by $\log_b b'$. In fact we shall take $b = e$, the base of natural logarithms, so that $\log_e x = \ln x$. By Lemma 3.9,

$$\ln(q_i/p_i) \leq (q_i/p_i) - 1,$$

with equality if and only if $q_i = p_i$. Since $\ln(q_i/p_i) = \ln(1/p_i) - \ln(1/q_i)$, we have

$$\begin{aligned}\sum_{i=1}^{m} p_i \ \ln(1/p_i) - \sum_{i=1}^{m} p_i \ \ln(1/q_i) &= \sum_{i=1}^{m} p_i \ \ln(q_i/p_i) \\ &\leq \sum_{i=1}^{m} p_i \ (q_i/p_i - 1) \\ &= \sum_{i=1}^{m} q_i - \sum_{i=1}^{m} p_i \\ &= 1 \ - \ 1 \ = \ 0,\end{aligned}$$

and equality holds if and only if $q_i = p_i$ for all i. □

Theorem 3.11

The entropy (uncertainty) of a distribution $\mathbf{p}$ on m symbols is at most $\log_b m$. The maximum value occurs if and only if all the symbols are equally probable.

Proof

In Theorem 3.10 take $q_i = 1/m$ $(1 \leq i \leq m)$. Then

$$H_b(\mathbf{p}) \leq \sum_{i=1}^{m} p_i \ \log_b m = \log_b m,$$

with equality if and only if $p_i = q_i = 1/m$ for all i $(1 \leq i \leq m)$. □

EXERCISES

3.10. Suppose a memoryless source emits three symbols a, b, c with probabilities $0.6, 0.3, 0.1$, and another memoryless source emits the same symbols, with probabilities $0.5, 0.3, 0.2$. Which source has the greater uncertainty? What probability distribution on the three symbols produces the greatest uncertainty?

3.11. Find the derivative $h'(x)$ of the function h whose graph is shown in Figure 3.1. Verify that the maximum value is at $x = \frac{1}{2}$ and that the tangent becomes vertical as x approaches 0 and 1.

3.12. Interpret the result of Exercise 3.9 in terms of 'uncertainty'.

3.5 Optimal codes – the fundamental theorems

We now return to the problem proposed in Section 3.2: how to find a UD b-ary code for a source $(S, \mathbf{p})$ such that the average word-length L is minimized. It turns out that the entropy $H_b(\mathbf{p})$ plays a crucial (and rather surprising) part.

Theorem 3.12

The average word-length of any UD b-ary code for the source $(S, \mathbf{p})$ satisfies

$$L \geq H_b(\mathbf{p}).$$

Proof

Denote by y_i the length of the codeword for the symbol s_i, $i = 1, 2, \ldots, m$. Then, as in Section 3.2, the Kraft-McMillan number for the code is

$$K = \frac{1}{b^{y_1}} + \frac{1}{b^{y_2}} + \cdots + \frac{1}{b^{y_m}}.$$

Put $q_i = 1/(Kb^{y_i})$ so that $\mathbf{q} = [q_1, q_2, \ldots, q_m]$ is a probability distribution. Applying the Comparison Theorem 3.10 to $\mathbf{p}$ and $\mathbf{q}$ we have

$$H_b(\mathbf{p}) = \sum_{i=1}^{m} p_i \log_b(1/p_i) \leq \sum_{i=1}^{m} p_i \log_b(1/q_i).$$

Since $1/q_i = Kb^{y_i}$, we have $\log_b(1/q_i) = \log_b K + y_i$, and so

$$H_b(\mathbf{p}) \leq \sum_{i=1}^{m} p_i(\log_b K + y_i) = \log_b K + \sum_{i=1}^{m} p_i y_i = \log_b K + L.$$

Since the code is UD, we have $K \leq 1$, so $\log_b K \leq 0$ and the result is proved. □

The obvious question is: how close to the lower bound $H_b(\mathbf{p})$ can L be? In fact it is easy to see that the bound cannot be attained in many cases, for numerical reasons. The last line of the proof of Theorem 3.12 shows that if $H_b(\mathbf{p}) = L$ then $\log_b K = 0$, so $K = 1$. Furthermore, it follows from Theorem 3.10 that equality can hold only if $p_i = q_i = 1/Kb^{y_i}$ for all i. In other words, for equality we require

$$b^{y_i} = 1/p_i \quad \text{for all } i.$$

However, in general $1/p_i$ is not an integral power of b. For example, in the binary case we can only achieve equality if all the probabilities belong to the set $\{1/2, 1/4, 1/8, 1/16, \ldots\}$.

The preceding argument suggests that we can try to construct a code with average word-length close to L by choosing b^{y_i} to be as close as possible to $1/p_i$. That is,

$$y_i \text{ is the least positive integer such that } b^{y_i} \geq 1/p_i.$$

This is sometimes called the *Shannon-Fano (SF) rule.* The resulting value of K is

$$K = \frac{1}{b^{y_1}} + \frac{1}{b^{y_2}} + \cdots + \frac{1}{b^{y_m}} \leq p_1 + p_2 + \cdots + p_m = 1,$$

and so (by Theorem 2.9) a PF code with these parameters does exist.

The next theorem shows that the average word-length L of such a code is reasonably close to $H_b(\mathbf{p})$.

Theorem 3.13

There is a PF b-ary code for a source with probability distribution $\mathbf{p}$ that satisfies the inequality

$$L < H_b(\mathbf{p}) + 1.$$

Proof

Let y_i $(i = 1, 2, \ldots, m)$ be the integer defined by the SF rule, so that a PF code with these parameters exists.

We shall show that $L < H_b(\mathbf{p}) + 1$ for this code. Since y_i is the *least* integer such that $b^{y_i} \geq 1/p_i$, we have $b^{y_i - 1} < 1/p_i$. That is, $y_i - 1 < \log_b(1/p_i)$. Hence, using the fact that $\sum p_i = 1$, we have

$$L = \sum p_i y_i < \sum p_i(1 + \log_b(1/p_i)) = 1 + H_b(\mathbf{p}).$$

□

Example 3.14

A source emits five symbols with probability distribution

$$\mathbf{p} = [0.4, 0.2, 0.2, 0.1, 0.1].$$

What is the entropy $H(\mathbf{p})$? Construct a binary code for this source with average word-length less than $H(\mathbf{p}) + 1$.

Solution Here we use the convention that $H(\mathbf{p})$ denotes entropy with respect to the base $b = 2$. A simple calculation shows that $H(\mathbf{p}) \approx 2.12$. The SF rule tells us to choose y_1 to be the least integer such that

$$2^{y_1} \geq 1/0.4 = 2.5,$$

that is, $y_1 = 2$. Similarly

$$y_2 = 3,\ y_3 = 3,\ y_4 = 4,\ y_5 = 4.$$

Thus the parameters required are $n_2 = 1, n_3 = 2, n_4 = 2$. Using the tree method, it is easy to construct a PF code with these parameters, for example:

$$00,\ 010,\ 011,\ 1000,\ 1001.$$

The average length is

$$L = 0.4 \times 2 + 0.2 \times 3 + 0.2 \times 3 + 0.1 \times 4 + 0.1 \times 4 = 2.8,$$

which (as guaranteed by Theorem 3.13) is less than $H(\mathbf{p}) + 1 \approx 3.12$.

EXERCISES

3.13. Use the Shannon-Fano coding rule to construct a PF binary code for a source with probability distribution

$$\mathbf{p} = [0.25, 0.10, 0.15, 0.05, 0.20, 0.25].$$

Find its average word-length L, and verify that L lies between $H(\mathbf{p})$ and $H(\mathbf{p}) + 1$.

3.14. Use the SF rule to construct a PF ternary code for a source with probability distribution $[0.5, 0.3, 0.2]$. Show, by constructing a better code, that this code is not optimal.

3.6 Huffman's rule

Recall that if a UD code exists, then it is possible to construct a PF code with the same parameters. Hence we can confine our search for optimal codes to codes that have the PF property.

For many purposes the Shannon-Fano rule produces a satisfactory code, but in general it does not give an optimal code. *Huffman's rule*, described below, is guaranteed to give an optimal code. (For simplicity, we shall describe the binary case only; a similar rule works when $b > 2$.)

Lemma 3.15

An optimal PF code $c : S \to \mathbb{B}^*$ for a source $(S, \mathbf{p})$ has the properties

(1) if the codeword $c(s')$ is longer than $c(s)$ then $p_s \geq p_{s'}$;

(2) among the codewords of maximum length there are two of the form $x0$ and $x1$, for some $x \in \mathbb{B}^*$.

Proof

(1) Suppose the length of $w = c(s)$ is α and the length of $w' = c(s')$ is α'. Let c^* be the code obtained by defining $c^*(s) = w'$ and $c^*(s') = w$ (clearly this is PF). Then the average word-lengths $L(c)$ and $L(c^*)$ satisfy

$$L(c^*) - L(c) = (p_s\alpha' + p_{s'}\alpha) - (p_s\alpha + p_{s'}\alpha') = (p_s - p_{s'})(\alpha' - \alpha).$$

Since c is optimal, this must be non-negative. Hence if $\alpha' > \alpha$ then $p_s \geq p_{s'}$.
(2) If no two words of maximum length have the form stated, then deleting the last bit from all codewords of maximum length would produce a better code that still has the PF property. □

Huffman's rule employs two constructions based on these properties (see Figure 3.3).

H1 Given a source $(S, \mathbf{p})$, let s' and s'' be two symbols with the smallest probabilities. Construct a new source $(S^*, \mathbf{p}^*)$ by replacing s' and s'' by a single symbol s^*, with probability $p^*_{s*} = p_{s'} + p_{s''}$. All other symbols have unchanged probabilities.

H2 If we are given a PF binary code h^* for $(S^*, \mathbf{p}^*)$, with $h^*(s^*) = w$, then a PF binary code h for $(S, \mathbf{p})$ is defined by the rules $h(s') = w0$, $h(s'') = w1$, and $h(u) = h^*(u)$ for all $u \neq s', s''$.

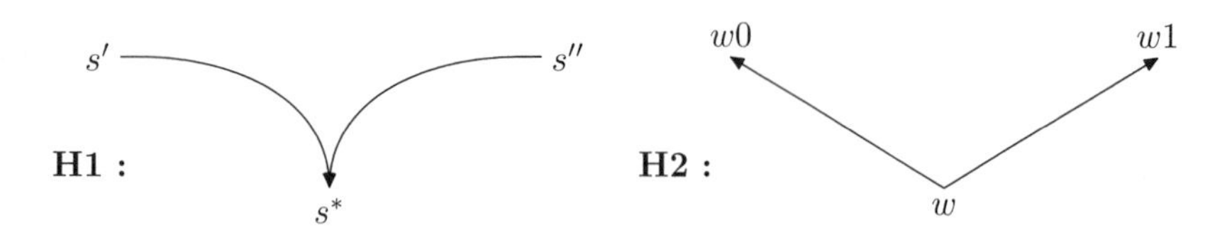

Figure 3.3 The two parts of Huffman's rule

Suppose we are given a source with m symbols. The rule **H1** can be used to construct a sequence of sources each with one symbol less than the previous one, so the process stops at the mth source, which has one symbol. The optimal code for the last source is the trivial one which assigns the empty word to the single symbol. Working backwards, **H2** can be used to construct codes for each of the sources in the sequence. The optimality of the resulting codes will be proved in the next section.

Example 3.16

Use Huffman's rule to construct an optimal code for a source with distribution $\mathbf{p} = [0.4, 0.2, 0.2, 0.1, 0.1]$.

Solution Starting with $\mathbf{p}^{(1)} = \mathbf{p}$ and defining $\mathbf{p}^{(i+1)} = \mathbf{p}^{(i)*}$ for $i = 1, 2, 3, 4$, the rule **H1** produces the sequence of sources shown in Figure 3.4.

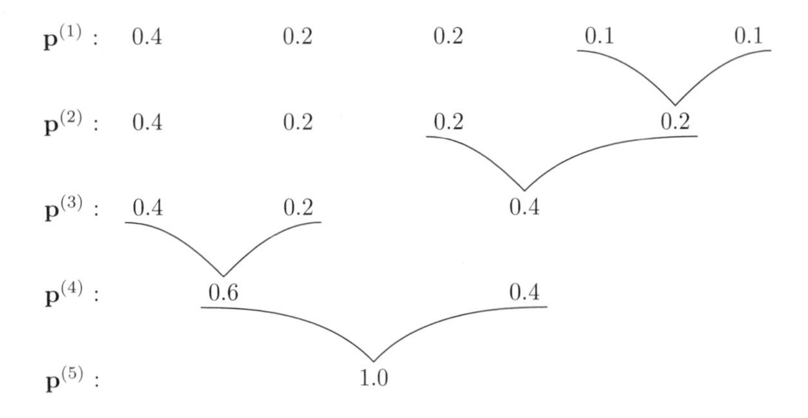

Figure 3.4 Application of rule **H1**

In order to construct the Huffman code we use **H2**, starting from the code that assigns the empty word to the single symbol in the last line. The process is shown in Figure 3.5. The codewords are

$$00, 01, 10, 110, 111.$$

For this source the average word-length of the Huffman code is

$$L_{opt} = (0.4 + 0.2 + 0.2) \times 2 \; + \; (0.1 + 0.1) \times 3 = 2.2.$$

The entropy $H = H(\mathbf{p})$ is 2.12 approximately. In Example 3.14 the SF rule was applied to this source, and a code with average word-length $L_{SF} = 2.8$

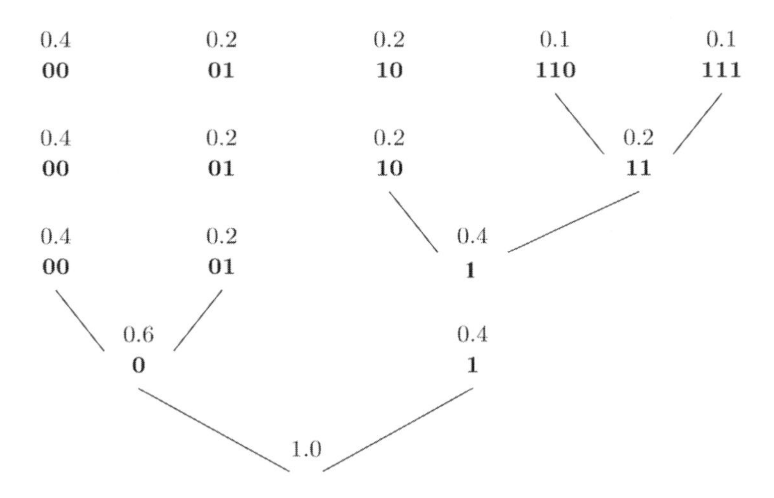

Figure 3.5 Application of rule **H2**

was obtained. The relationship between these results and the bounds obtained in Section 3.5 can be summarized as:

$$H \; < \; L_{opt} \; < \; L_{SF} \; < \; H+1.$$

EXERCISES

3.15. Construct an optimal code for the source in Exercise 3.13 using Huffman's rule, and find its average word-length.

3.16. Consider a source with probability distribution $[0.4, 0.3, 0.2, 0.1]$. Compare the average word-length of the binary code obtained from the SF rule with the average word-length of the optimal one, obtained by Huffman's rule.

3.17. Suppose that seven symbols are emitted by a source, with probability distribution

$$\mathbf{p} \; = \; [0.2, 0.2, 0.2, 0.1, 0.1, 0.1, 0.1].$$

What is the entropy of the source?

Use the Shannon-Fano rule to construct a prefix-free code for this source. Find its average word-length L, and verify that Theorems 3.10 and 3.12 hold. Use Huffman's rule to construct an optimal binary code for this source.

3.18. Find the Huffman code for a source with probability distribution

$$[0.4,\ 0.3,\ 0.1,\ 0.1,\ 0.06,\ 0.04].$$

3.7 Optimality of Huffman codes

Lemma 3.17

Let h and h^* be defined as in the construction **H2**. Then the average word-lengths of h and h^* satisfy

$$L(h) = L(h^*) + p^*_{s^*}.$$

Proof

Suppose that the length of $h^*(s^*)$ is β. Then in the code h the symbols s' and s'' are assigned codewords of length $\beta + 1$. Hence

$$L(h) - L(h^*) = (p_{s'} + p_{s''})(\beta + 1) - p^*_{s^*}\beta.$$

Since $p_{s'} + p_{s''} = p^*_{s^*}$, this expression reduces to $p^*_{s^*}$. □

Theorem 3.18

If h^* is optimal for $(S^*, \mathbf{p}^*)$ then h is optimal for $(S, \mathbf{p})$.

Proof

We prove the contrapositive: that is, if h is not optimal then h^* is not optimal.

Suppose h is not optimal, and let c be an optimal code for $(S, \mathbf{p})$. It follows from part (ii) of Lemma 3.15 that there are symbols t', t'' to which c assigns codewords $x0$ and $x1$ of maximal length. Suppose first that the pair (t', t'') is disjoint from the pair (s', s'') involved in the construction of h. Then

$$c(t') = x0,\ c(t'') = x1,\ c(s') = y,\ c(s'') = z,$$

where the codewords have length μ, μ, γ, δ, say.

Define a code c^* for $(S^*, \mathbf{p}^*)$ by

$$c^*(t') = y,\ c^*(t'') = z,\ c^*(s^*) = x.$$

Here the lengths of the codewords are $\gamma, \delta, \mu - 1$. Hence

$$L(c) - L(c^*) = (p_{t'} + p_{t''})\mu + p_{s'}\gamma + p_{s''}\delta - p_{t'}\gamma - p_{t''}\delta - p^*_{s^*}(\mu - 1).$$

Since $p^*_{s^*} = p_{s'} + p_{s''}$, this is can be rewritten as

$$p^*_{s^*} + (p_{t'} - p_{s'})(\mu - \gamma) + (p_{t''} - p_{s''})(\mu - \delta).$$

Now $x0$ and $x1$ are codewords of maximum length, so $\mu \geq \gamma$ and $\mu \geq \delta$. Also, since s', s'' are symbols with smallest probability, $p_{t'} \geq p_{s'}$ and $p_{t''} \geq p_{s''}$. Hence

$$L(c) - L(c^*) \geq p^*_{s^*}.$$

We know that $L(h) = L(h^*) + p^*_{s^*}$ (Lemma 3.17), and since h is assumed to be not optimal, $L(c) < L(h)$. Thus

$$L(c^*) \leq L(c) - p^*_{s^*} < L(h) - p^*_{s^*} = L(h^*),$$

and h^* is not optimal for $(S^*, \mathbf{p}^*)$.

Note that if (t', t'') and (s', s'') overlap, minor changes are needed. □

EXERCISES

3.19. Show that the sum of the 'new' probabilities $p^*_{s^*}$ obtained in the application of rule **H1** is equal to L, the average length of the Huffman code. (This result shows that L can be found using only the first part of Huffman's rule, without having to use **H2** to find the code explicitly.)

3.20. Suppose that $\mathbf{p}$ is a source with N symbols and and $\ell(C)$ is the *total* length of the codewords in an optimal binary code C for $\mathbf{p}$. Show that if $C, C^{(1)}, \ldots C^{(N-3)}, C^{(N-2)}$ is the sequence of codes constructed by the Huffman rules, then

$$\ell(C^{(N-i-1)}) \leq \ell(C^{(N-i)}) + (i+1).$$

Deduce that

$$\ell(C) \leq \frac{1}{2}(N^2 + N - 2).$$

3.21. Find the optimal encoding of the source defined in Exercise 3.16 by a *ternary* code - that is, using the symbols $\{0, 1, 2\}$. Justify the assertion that the code is optimal. [It is not necessary to use Huffman's rule for this exercise.]

3.22. Let m be a positive integer and let $\mathbf{p}_m$ denote the probability distribution on a set of m symbols in which each symbol is equiprobable:

$$\mathbf{p}_m = [1/m, 1/m, \ \ldots \ , 1/m].$$

For each m in the range $1 \leq m \leq 8$, find the optimal binary code for the distribution $\mathbf{p}_m$, and its average word-length. On the basis of your results, make a conjecture about the solution for a general m and illustrate it in the case $m = 400$. Describe the relationship between the entropy H and the average word-length L.

Further reading for Chapter 3

In order to define the concept of a source in full generality it is necessary to use some quite sophisticated probability theory. A good account is given by Goldie and Pinch [**3.2**]. The Bayesian approach to the subject is comprehensively covered by MacKay [**3.4**].

The concept of entropy was first studied in theoretical physics, specifically thermodynamics. The connections with information theory were fully explored in the paper by Shannon [**3.5**], which effectively laid the foundations for the whole subject. The standard text by Cover and Thomas [**3.1**] is useful. The proof that the entropy function is essentially the only function satisfying a number of conditions that we associate with the idea of uncertainty is given in Appendix 1 of Welsh's book [**3.6**].

Huffman's algorithm [**3.3**] was discovered when the subject was still in its infancy.

3.1 T.M. Cover and J.A. Thomas. *Elements of Information Theory.* Wiley (Second edition, 2006).

3.2 C.M. Goldie and R.G.E. Pinch. *Communication Theory.* Cambridge University Press (1991).

3.3 D.A. Huffman. A method for the construction of minimum redundancy codes. *Proc. IRE* 40 (1952) 1098-1011.

3.4 D.J.C. MacKay. *Information Theory, Inference, and Learning Algorithms.* Cambridge University Press (2003).

3.5 C.E. Shannon. A mathematical theory of communication. *Bell System Tech. J.* 27 (1948) 379-423, 623-656.

3.6 D.J.A. Welsh. *Codes and Cryptography.* Oxford University Press (1988).

4
Data compression

4.1 Coding in blocks

We have formalized the idea of a code as a function $c : S \to T^*$ that replaces symbols in an alphabet S by strings of symbols in an alphabet T. Although we have chosen to regard the elements of S as single objects, such as the letters of a natural alphabet, there is no logical need for this restriction. In practice, it is often useful to split the stream of symbols emitted by a source into blocks (disjoint strings of symbols), and to regard the blocks themselves as symbols.

For example, suppose a source emits a stream of letters, each of which is x, y, or z, so that a typical stream would be

$$yzxxzyxzyxzzyxxzyzxzzyyy \cdots .$$

Here the alphabet S is the set $\{x, y, z\}$. On the other hand, we could split the stream into blocks of length 2:

$$yz\ xx\ zy\ xz\ yx\ zz\ yx\ xz\ yz\ xz\ zy\ yy\ \cdots .$$

Now the alphabet is the set S^2 of ordered pairs

$$S^2 = \{xx, xy, xz, yx, yy, yz, zx, zy, zz\}.$$

In Chapter 3 we considered the problem of coding a source $(S, \mathbf{p})$ so that the average word-length L is as small as possible. We found that the entropy $H_b(\mathbf{p})$ is a lower bound for L, but this bound is rarely attained. We shall now explain how coding in blocks enables us to approach the lower bound as closely as we please.

N. L. Biggs, *An Introduction to Information Communication and Cryptography*,
DOI: 10.1007/978-1-84800-273-9_4,

Example 4.1

Consider a scanner processing a black-and-white document. Suppose it emits W (white pixel) and B (black pixel) with probabilities $p_W = 0.9$, $p_B = 0.1$. Calculate the entropy of this distribution. Assuming that the source is memoryless, find the best binary codes for it using blocks of size 1, 2, and 3.

Solution The entropy is

$$H(\mathbf{p}) = 0.9\log_2(1/0.9) + 0.1\log_2(1/0.1) \approx 0.469.$$

Using blocks of size 1, the best way to code the source is the obvious one, $W \mapsto 0$, $B \mapsto 1$, because any other code must use a codeword of length greater than 1. This code has average word-length $L_1 = 1$, which is much worse than the lower bound 0.469. A document containing N pixels is encoded by a string of N bits, although theoretically speaking, the amount of information is equivalent to only $0.469N$ bits.

Consider next what happens when we split the message into blocks of size 2. Now we have a source with four 'symbols', and the probabilities are:

$$\begin{array}{cccc} WW & WB & BW & BB \\ 0.81 & 0.09 & 0.09 & 0.01 \end{array}.$$

What is the optimal binary code for this source? A simple application of Huffman's rule gives the code

$$WW \mapsto 0, \quad WB \mapsto 10, \quad BW \mapsto 110, \quad BB \mapsto 111.$$

The average length of this code is

$$L_2 = 0.81 \times 1 + 0.09 \times 2 + (0.09 + 0.01) \times 3 = 1.29.$$

Observe that a 'symbol' now represents *two pixels*. A document with N pixels will contain $N/2$ 'symbols', and will be encoded by a string with $L_2N/2 = 0.645N$ bits, approximately. Thus we have a significant improvement on our first attempt to achieve the theoretical lower bound of $0.469N$.

What happens if we use blocks of length 3? Now the source has eight 'symbols', and the probabilities are:

$$\begin{array}{cccccccc} WWW & WWB & WBW & WBB & BWW & BWB & BBW & BBB \\ 0.729 & 0.081 & 0.081 & 0.009 & 0.081 & 0.009 & 0.009 & 0.001 \end{array}.$$

Another application of Huffman's rule produces the code

$$WWW \mapsto 0,\ WWB \mapsto 100,\ WBW \mapsto 101,\ WBB \mapsto 11100,$$

$$BWW \mapsto 110,\ BWB \mapsto 11100,\ BBW \mapsto 11110,\ BBB \mapsto 11111,$$

for which the average length is $L_3 = 1.598$. Repeating the previous argument, we see that now a document containing N pixels can be encoded by a string of $L_3N/3 = 0.533N$ bits approximately, which is closer still to the theoretical lower bound of $0.469N$.

The technique described in the example is the basis of *data compression*. In the rest of this chapter we shall explore its theoretical foundations, and describe some of the coding rules that can be used to implement it.

EXERCISES

4.1. How many words of length ℓ can be formed from an alphabet with r symbols? A message using an alphabet with r symbols has length k. If it contains all possible words of length ℓ as sub-strings (not necessarily disjoint), what is the smallest possible value of k?

4.2. Construct a message that attains the minimum value found in the previous exercise, in the case $r = 3$, $\ell = 2$.

4.3. Consider a memoryless source that emits symbols A and B with probabilities $p_A = 0.8$, $p_B = 0.2$. Calculate the entropy for this source. Find binary Huffman codes for the associated sources using blocks of size 2 and 3, and calculate their average word-lengths L_2 and L_3. Can you estimate the limit of L_n/n as $n \to \infty$?

4.2 Distributions on product sets

Let $S' = \{s'_1, s'_2, \ldots, s'_m\}$, $S'' = \{s''_1, s''_2, \ldots, s''_n\}$, and let $Y = S' \times S''$ be the product set, containing the pairs of symbols $s'_i s''_j$. Let $\mathbf{p}$ be a probability distribution on Y, and denote the probability of $s'_i s''_j$ by p_{ij}.

Definition 4.2 (Marginal distributions, independence)

With the notation as above, let

$$p'_i = \sum_{j=1}^{n} p_{ij} \quad (i = 1, 2, \ldots, m), \qquad p''_j = \sum_{i=1}^{m} p_{ij} \quad (j = 1, 2, \ldots, n).$$

The distributions $\mathbf{p}'$ on S' and $\mathbf{p}''$ on S'' are known as the *marginal distributions* associated with $\mathbf{p}$. Clearly, p'_i is the probability that the first component is s'_i, and p''_j is the probability that the second component is s''_j. The distributions $\mathbf{p}'$ and $\mathbf{p}''$ are *independent* if $p_{ij} = p'_i p''_j$.

Example 4.3

Suppose that $S' = \{a, b\}$, $S'' = \{c, d\}$ and the distribution $\mathbf{p}$ is given by the following table. (For example, the entry in row a and column d is p_{ad}.)

	c	d
a	0.3	0.1
b	0.4	0.2

Find the marginal distributions $\mathbf{p}'$ and $\mathbf{p}''$. Are these distributions independent?

Solution We have

$$p'_a = p_{ac} + p_{ad} = 0.4, \qquad p'_b = p_{bc} + p_{bd} = 0.6,$$

$$p''_c = p_{ac} + p_{bc} = 0.7, \qquad p''_d = p_{ad} + p_{bd} = 0.3.$$

The distributions are not independent because (for example)

$$p_{ac} = 0.3 \qquad \text{whereas} \qquad p'_a p''_c = 0.28.$$

Theorem 4.4

The entropies of the distributions $\mathbf{p}$, $\mathbf{p}'$, $\mathbf{p}''$ satisfy

$$H(\mathbf{p}) \;\leq\; H(\mathbf{p}') + H(\mathbf{p}'').$$

Equality holds if and only if $\mathbf{p}'$ and $\mathbf{p}''$ are independent.

Proof

By definition

$$H(\mathbf{p}') + H(\mathbf{p}'') = \sum_i p'_i \log(1/p'_i) + \sum_j p''_j \log(1/p''_j).$$

Since $p'_i = \sum_j p_{ij}$ and $p''_j = \sum_i p_{ij}$, we have

$$\begin{aligned} H(\mathbf{p}') + H(\mathbf{p}'') &= \textstyle\sum_i \sum_j p_{ij} \log(1/p'_i) + \sum_j \sum_i p_{ij} \log(1/p''_j) \\ &= \textstyle\sum_{i,j} p_{ij} \; \log(1/p'_i p''_j). \end{aligned}$$

Consider two probability distributions defined on $S' \times S''$: the given distribution $\mathbf{p}$, and $\mathbf{q}$ defined by $q_{ij} = p'_i p''_j$. Applying the Comparison Theorem (3.10) it follows that

$$H(\mathbf{p}) \leq \sum_{i,j} p_{ij} \log(1/q_{ij}) = \sum_{i,j} p_{ij} \log(1/p'_i p''_j),$$

which is just $H(\mathbf{p}') + H(\mathbf{p}'')$, as shown above. The inequality is an equality if and only if $\mathbf{p} = \mathbf{q}$, that is, $p_{ij} = p'_i p''_j$, which is the condition that $\mathbf{p}'$ and $\mathbf{p}''$ are independent. □

Similar results hold for a product of more than two sets, and they can be proved quite simply by induction. We shall be mainly concerned with the case when all the sets are the same, that is, with the product S^r of $r \geq 2$ copies of a given set S. In that case an element of S^r is just a word of length r in the alphabet S.

Example 4.5

Suppose a source emits a stream of bits, and a number of observations suggest that the frequencies of blocks of length 2 are given by the following probability distribution $\mathbf{p}$ on $\mathbb{B}^2$. For example, $p_{01} = 0.4$ means that about 40 out of every 100 blocks are 01.

	0	1
0	0.1	0.4
1	0.4	0.1

Find the marginal distributions $\mathbf{p}'$ and $\mathbf{p}''$, calculate their entropies, and check that Theorem 4.4 holds.

Solution The marginal distributions are $\mathbf{p}' = [0.5, 0.5]$ and $\mathbf{p}'' = [0.5, 0.5]$. Thus $H(\mathbf{p}') = H(\mathbf{p}'') = h(0.5) = 1$. On the other hand,

$$H(\mathbf{p}) = 0.1 \log(1/0.1) + 0.4 \log(1/0.4) + 0.4 \log(1/0.4) + 0.1 \log(1/0.1) \approx 1.722.$$

Thus Theorem 4.4 holds.

In the example the stream has the property that each individual bit is equally likely to be 0 or 1, although the pairs of bits are not equally distributed. The marginal distributions $\mathbf{p}'$ and $\mathbf{p}''$ are the same, but they are not independent since, for example, $p_{00} = 0.1$ whereas $p'_0 p''_0 = 0.25$. In other words the source is not *memoryless* (Definition 3.2).

EXERCISES

4.4. Suppose that $X' = \{u, v, w\}$, $X'' = \{y, z\}$, and the distribution $\mathbf{p}$ on $X' \times X''$ is given by the table

	y	z
u	0.2	0.1
v	0.3	0.1
w	0.1	0.2

Are the marginal distributions $\mathbf{p}'$ and $\mathbf{p}''$ independent?

4.5. Verify that the entropies of the distributions defined in the previous exercise satisfy Theorem 4.4.

4.6. Suppose it is observed that, in a certain stream of bits, the frequencies of blocks of length 2 are given by the following probability distribution $\mathbf{p}$ on $\mathbb{B}^2$.

	0	1
0	0.35	0.15
1	0.15	0.35

Find the marginal distributions $\mathbf{p}'$ and $\mathbf{p}''$, calculate their entropies, and check that Theorem 4.4 holds.

4.3 Stationary sources

In Chapter 3 we considered the stream $\xi_1\xi_2\xi_3\cdots$ emitted by a source as a sequence of identically distributed random variables ξ_k $(k = 1, 2, 3\ldots)$, taking values in a set S.

Suppose we consider the stream emitted by a source as a stream of blocks, where each block is a random variable that takes values in the set S^r of strings of length r. For example, taking $r = 2$ we have a stream of random variables $\xi_{2k-1}\xi_{2k}$. We should like to define a probability distribution $\mathbf{p}^2$ on S^2 by the rule

$$\mathbf{p}^2(s_i s_j) = \Pr(\xi_{2k-1}\xi_{2k} = s_i s_j) = \Pr(\xi_{2k-1} = s_i,\ \xi_{2k} = s_j).$$

Clearly, in order to do this it must be assumed that the probability of emitting the given pair of symbols $s_i s_j$ does not depend on k, the position in the stream. The general form of this condition is as follows.

Definition 4.6 (Stationary source)

A source emitting a stream $\xi_1\xi_2\xi_3\cdots$ is *stationary* if, for any positive integers $\ell_1, \ell_2, \ldots, \ell_r$, the probability

$$\Pr(\xi_{k+\ell_1} = x_1,\ \xi_{k+\ell_2} = x_2,\ \ldots\ , \xi_{k+\ell_r} = x_r)$$

depends only on the string $x_1x_2\ldots x_r$, not on k.

Although there are many situations where it is reasonable to assume that the condition holds in its general form, in practice we can only check its validity in a few cases. Usually we consider *consecutive* symbols, that is, the case $\ell_1 = 1$, $\ell_2 = 2, \ldots, \ell_r = r$ of the definition. Then the definition implies that for a stationary source we have probability distributions $\mathbf{p}^r$, defined on S^r for $r \geq 1$ by the rule

$$\mathbf{p}^r(x_1x_2\ldots x_r) \;=\; \Pr(\xi_{k+1} = x_1,\ \xi_{k+2} = x_2,\ \ldots\ , \xi_{k+r} = x_r) \quad \text{for all } k.$$

When $r = 1$ the definition reduces to our standard assumption (Section 3.1) that all the random variables ξ_k have the same distribution $\mathbf{p}^1$.

Roughly speaking, stationarity means that the probability that a given word (string of consecutive symbols) will appear on 'page 1' of the message is the same as the probability that it will appear on 'page 99', or any other 'page'. In practice, we can often check this property experimentally for strings of a certain length r. Then we can use $\mathbf{p}^r$ to determine the marginal probability distributions $\mathbf{p}^s$ for $1 \leq s < r$, using the addition law of probability. For instance the distribution on strings of length $r-1$ is related to the distribution on strings of length r by the equations

$$\mathbf{p}^{r-1}(x_1x_2\cdots x_{r-1}) \;=\; \sum_{s\in S}\mathbf{p}^r(x_1x_2\cdots x_{r-1}s) \;=\; \sum_{s\in S}\mathbf{p}^r(sx_1\cdots x_{r-1}).$$

We have already encountered these equations in the case $r = 2$. Note that, for a stationary source, it follows that the two marginal distributions associated with $\mathbf{p}^2$ are the same.

A *memoryless* source is a very special case of a stationary source. In that case, each $\mathbf{p}^\mathbf{r}$ is simply determined by $\mathbf{p}^1$:

$$\mathbf{p}^r(x_1x_2\ldots x_r) = \mathbf{p}^1(x_1)\mathbf{p}^1(x_2)\cdots\mathbf{p}^1(x_r).$$

Example 4.7

Consider a stationary source emitting symbols from the alphabet $S = \{a, b, c\}$, with the probability distribution $\mathbf{p}^2$ on S^2 defined by the following table.

	a	b	c
a	0.39	0.17	0.04
b	0.15	0.11	0.04
c	0.06	0.02	0.02

Find the corresponding probability distribution $\mathbf{p}^1$ on S. Is this source memoryless? What is the relationship between $H(\mathbf{p}^2)$ and $H(\mathbf{p}^1)$?

Solution According to the addition law,

$$\mathbf{p}^1(a) \;=\; \mathbf{p}^2(aa) + \mathbf{p}^2(ab) + \mathbf{p}^2(ac) \;=\; 0.6.$$

Similarly $\mathbf{p}^1(b) = 0.3$ and $\mathbf{p}^1(c) = 0.1$. The source is not memoryless, since (for example)

$$\mathbf{p}^2(aa) = 0.39 \qquad \text{whereas} \qquad \mathbf{p}^1(a)\mathbf{p}^1(a) = 0.36.$$

The entropy of $\mathbf{p}^1$ is

$$0.6\log(1/0.6) + 0.3\log(1/0.3) + 0.1\log(1/0.1) \;\approx\; 1.295.$$

A similar calculation gives the entropy of $\mathbf{p}^2$ as 2.520 approximately. Thus $H(\mathbf{p}^2) < H(\mathbf{p}^1) + H(\mathbf{p}^1)$, in agreement with Theorem 4.4.

EXERCISES

4.7. Suppose that a stationary source emits symbols from the alphabet $S = \{a, b, c, d\}$ and the probability distribution $\mathbf{p}^2$ is given by the table

	a	b	c	d
a	0.14	0.17	0.04	0.05
b	0.15	0.10	0.04	0.01
c	0.05	0.02	0.10	0.03
d	0.06	0.01	0.02	0.01

What is the distribution $\mathbf{p}^1$? Is this a memoryless source?

4.8. Suppose that a source emits a stream of bits, and observations suggest that the frequencies of the blocks of length 2 are given by the following probability distribution on $\mathbb{B}^2$.

	0	1
0	0.2	0.3
1	0.3	0.2

In Section 4.2 we noted that this is not a memoryless source. Show that it is not necessarily stationary, by constructing a probability distribution $\mathbf{p}^3$ on $\mathbb{B}^3$ that can vary in time, but is nevertheless consistent with the observations.

4.4 Coding a stationary source

Suppose we regard the stream emitted by a source as a stream of blocks of length r. If the source is stationary then, for each $r \geq 1$, there is an associated probability distribution $\mathbf{p}^r$, and its entropy $H(\mathbf{p}^r)$ is defined. This represents the uncertainty of the stream, *per block* of r symbols.

In order to apply the fundamental theorems relating entropy and average word-length to a stationary source, we must begin with a definition of the entropy of such a source. For blocks of r symbols, the uncertainty *per symbol* is $H(\mathbf{p}^r)/r$, so the uncertainties, measured in bits per symbol, associated with the distributions $\mathbf{p}^r$ $(r \geq 1)$ are

$$H(\mathbf{p}^1), \quad \frac{H(\mathbf{p}^2)}{2}, \quad \frac{H(\mathbf{p}^3)}{3}, \quad \ldots, \quad \frac{H(\mathbf{p}^r)}{r}, \quad \ldots \quad .$$

This set of real numbers has a lower bound 0, since the entropy of any distribution is non-negative. Hence, by a fundamental property of the real numbers, it has a *greatest* lower bound. This is a number H with the properties (i) H is a lower bound; (ii) no number greater than H is a lower bound. It is customary to refer to the greatest lower bound of a set X as the *infimum* of the set, written $\inf X$.

Definition 4.8 (Entropy of a stationary source)

The *entropy* H of a stationary source with probability distributions $\mathbf{p}^r$ is the infimum of the numbers

$$H(\mathbf{p}^r)/r \qquad r = 1, 2, 3, \ldots \ .$$

This definition is consistent with the definition of the entropy of a memoryless source (Section 3.2.).

Theorem 4.9

If a stationary source is memoryless, its entropy H is equal to $H(\mathbf{p}^1)$.

Proof

If the source is memoryless then, by Theorem 4.4, $H(\mathbf{p}^2) = 2H(\mathbf{p}^1)$. In fact, a simple induction proof shows that $H(\mathbf{p}^r) = rH(\mathbf{p}^1)$ for all r, and so all the terms $H(\mathbf{p}^r)/r$ are equal to $H(\mathbf{p}^1)$. □

In general, the behaviour of the numbers $H(\mathbf{p}^r)/r$ is constrained by the following consequence of Theorem 4.4.

Lemma 4.10

Suppose n is a multiple of r. Then

$$\frac{H(\mathbf{p}^n)}{n} \leq \frac{H(\mathbf{p}^r)}{r}.$$

Proof

Suppose first that $n = 2r$. For any $\ell \geq 1$, the distribution of $\xi_{\ell+1}\xi_{\ell+2}\ldots\xi_{\ell+r}$ and $\xi_{\ell+r+1}\xi_{\ell+r+2}\ldots\xi_{\ell+2r}$ are $\mathbf{p}^r$, by the stationary property. The distribution of $\xi_{\ell+1}\ldots\xi_{\ell+r}\xi_{\ell+r+1}\ldots\xi_{\ell+2r}$ is is $\mathbf{p}^{2r}$. Hence, by Theorem 4.4,

$$\frac{H(\mathbf{p}^{2r})}{2r} \leq \frac{1}{2r}\left(H(\mathbf{p}^r) + H(\mathbf{p}^r)\right) = \frac{H(\mathbf{p}^r)}{r}.$$

A straightforward generalization of the preceding argument shows that, for any $q \geq 2$,

$$H(\mathbf{p}^{qr}) \leq H(\mathbf{p}^{(q-1)r}) + H(\mathbf{p}^r).$$

Hence the result follows by induction on q. □

In practice, it is not feasible to determine the values $H(\mathbf{p^r})/r$ for more than a few small values of r. We take the least of these values as an approximation to the actual lower bound H. In theory it is possible to find r so that the approximation is as close as we please, and this leads to a fundamental result: it is possible to construct codes with average word-length (in bits per symbol) arbitrarily close to H.

Theorem 4.11 (The coding theorem for stationary sources)

Suppose we have a stationary source emitting symbols from an alphabet S, with entropy H. Then, given $\epsilon > 0$, there is a positive integer n and a prefix-free binary code for $(S^n, \mathbf{p}^n)$ for which the average word-length L_n satisfies

$$\frac{L_n}{n} < H + \epsilon.$$

Proof

Since H is the *greatest* lower bound for the numbers $H(\mathbf{p}^r)/r$, there is an r such that

$$\frac{H(\mathbf{p}^r)}{r} < H + \frac{\epsilon}{2}.$$

Choose an integer q such that $q > 2/\epsilon r$ and let $n = qr$, so that $1/n < \epsilon/2$. By Lemma 4.10, $H(\mathbf{p}^n)/n \leq H(\mathbf{p}^r)/r$.

It follows from Theorem 3.13 there is a prefix-free binary code for $\mathbf{p}^n$ with average word-length $L_n < H(\mathbf{p}^n) + 1$. Hence

$$\frac{L_n}{n} < \frac{H(\mathbf{p}^n)}{n} + \frac{1}{n} < \frac{H(\mathbf{p}^r)}{r} + \frac{\epsilon}{2} < H + \epsilon.$$

□

EXERCISES

4.9. Show that the entropy of the source described in Exercise 4.7 does not exceed 1.26 approximately.

4.10. Suppose that a stationary source emits symbols from an alphabet S according to the probability distributions $\mathbf{p}^r$ on S^r $(r \geq 1)$. Show that

$$H(\mathbf{p}^{a+b}) \leq H(\mathbf{p}^a) + H(\mathbf{p}^b) \quad \text{for all } a, b \geq 1.$$

4.11. Let n and r be such that $n = qr + s$, where $0 \leq s < r$. Show that

$$\frac{H(\mathbf{p}^n)}{n} \leq \frac{H(\mathbf{p}^r)}{r}\left(1 - \frac{s}{n}\right) + \frac{H(\mathbf{p}^s)}{n}.$$

Deduce that

$$\lim_{n\to\infty} \frac{H(\mathbf{p}^n)}{n} = H,$$

where H is the entropy of the source.

[Hint. Given $\epsilon > 0$, choose r such that $H(\mathbf{p}^r)/r < H + \frac{1}{2}\epsilon$, and let K be the maximum value of $H(\mathbf{p}^s)$ for $0 \leq s < r$. Hence find a value n_0 such that, for all $n \geq n_0$, $H \leq H(\mathbf{p}^n)/n < H + \epsilon$.]

4.5 Algorithms for data compression

We can now see clearly how the technique described in Section 4.1 enables data to be compressed. Suppose a file of N bits is produced by a stationary source for which the entropy is H. Then, according to Theorem 4.11, by coding the file in blocks of sufficient length, we can reduce the size of the file to $(H+\epsilon)N$ bits, where ϵ is as small as we please.

Huffman's rule is not well-suited to this method in practice. It requires the construction of codewords for all the blocks before any coding can be done, and if the required block-length is n, the set of codewords will have size 2^n.

Example 4.12

Suppose the source is a scan of a black-and-white line drawing, such as the graph of a function (Figure 4.1). Assuming that about 1% of the pixels are black, give an upper bound for the entropy, and estimate the number of codewords that would be required to compress to within 25% of this value.

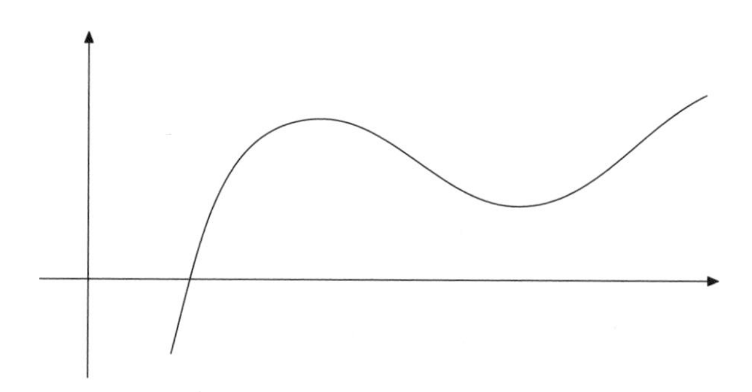

Figure 4.1 A drawing that defines a source of black and white pixels

Solution It is given that $H(\mathbf{p}^1) = h(0.01) \approx 0.08$ and so the entropy H of the source is such that $H \leq 0.08$. Hence, if we wish to compress to within 25% of the optimum value, we must choose ϵ in Theorem 4.11 to be no larger than 0.02, approximately. To achieve this we must use blocks of length $n \geq 2/\epsilon \approx 100$. Thus 2^{100} codewords may be required.

Another weakness of Huffman's rule is that it assumes that the characteristics of the source are known in advance. Often we wish to update the probabilities in course of processing, so that they better reflect what we know about the source. In fact, Huffman's rule can be modified to cope with this situation, but the modification is not simple.

In the light of these considerations there is clearly a need for alternative methods of coding with the following properties:

- the codeword for a given block can be calculated in isolation, without the need to find all the codewords;
- the codewords can be updated 'on-line' (this is known as *adaptive coding*).

In the following sections we shall describe two such methods. The first one, known as *arithmetic coding*, is based on the idea of assigning short codewords to symbols that occur with high frequencies. Although it is not strictly optimal, it can be proved to be close-to-optimal, and it can be implemented so that the two requirements are satisfied. The second technique, known as *dictionary coding*, is based on a more heuristic approach, but nevertheless it has been found to work well in practice.

EXERCISES

4.12. Suppose the source is scanning black-and-white diagrams that contain about 10% black pixels. Give an upper bound for the entropy, and estimate the size of the code that would be required to compress to within 25% of this value.

4.13. Consider a memoryless source emitting symbols from the alphabet $S = \{A, B\}$ with probabilities $p_A = x$, $p_B = 1 - x$. Let $L_2(x)$ denote the average word-length of the associated Huffman code on S^2. Draw the graph of $L_2(x)$ for $0 \leq x \leq 1$, noting any significant features.

4.6 Using numbers as codewords

The technique known as *arithmetic coding* is based on a correspondence between binary words (elements of $\mathbb{B}^*$) and fractions (elements of the set $\mathbb{Q}$ of *rational* numbers).

For any word $z_1 z_2 \cdots z_n$ in $\mathbb{B}^n$ there is a rational number

$$\frac{z_1}{2} + \frac{z_2}{2^2} + \cdots + \frac{z_n}{2^n}.$$

This number lies between 0 and 1, and it is denoted by $0.z_1 z_2 \ldots z_n$ in *binary notation*. Thus, for each integer $n \geq 1$ we have a function $\mathbb{B}^n \to \mathbb{Q}$, defined by

$$z_1 z_2 \cdots z_n \mapsto 0.z_1 z_2 \ldots z_n.$$

Example 4.13

Write down the rational numbers corresponding to the words 0110101 and 1010000.

Solution Here $n = 7$. The word 0110101 corresponds to the rational number

$$0.0110101 = \frac{1}{4} + \frac{1}{8} + \frac{1}{32} + \frac{1}{128} = \frac{53}{128},$$

or 0.4140625 in the usual (*decimal*) notation.

Similarly, the word 1010000 corresponds to the rational number

$$0.1010000 = \frac{1}{2} + \frac{1}{8} = \frac{5}{8},$$

or 0.625 in decimal notation. Note that the final 0's are normally omitted, so the binary form would be written as 0.101. But when it is important to record the value of n, the final 0's in the binary notation must be shown.

In arithmetic coding we use rational numbers to define codewords representing strings of symbols. The aim is to obtain codes that are close to optimal. As usual we do this by trying to ensure that a string X with high probability is represented by a codeword $c(X)$ with small length n. The codewords will therefore be defined in terms of the probabilities.

We begin with the set $\mathbb{R}$ of *real* numbers, because the probability of an event is usually allowed to be a real number, rather than a rational number. We shall choose the codeword $c(X)$ to be a number in a suitable interval of the form

$$[a, a + P) = \{r \in \mathbb{R} \mid a \leq r < a + P\},$$

where a is a real number and P is the probability of X. The next theorem explains how we can pick a rational number of the form $0.z_1z_2 \ldots z_n$, however small the length P of the interval.

Theorem 4.14

Suppose a and P are such that $0 \leq a < a + P \leq 1$. Let n be any integer such that $2^n > 1/P$. Then there is a word $z_1z_2 \ldots z_n \in \mathbb{B}^n$ such that

$$0.z_1z_2 \ldots z_n \; \in \; [a, a + P).$$

Proof

The interval $[0, 1)$ is partitioned into 2^n disjoint intervals J_i of length $1/2^n$, where

$$J_i \; = \; \left[\frac{i-1}{2^n}, \frac{i}{2^n}\right) \qquad i = 1, 2, \ldots, 2^n.$$

The given number a is in exactly one of these intervals, say J_c (Figure 4.2).

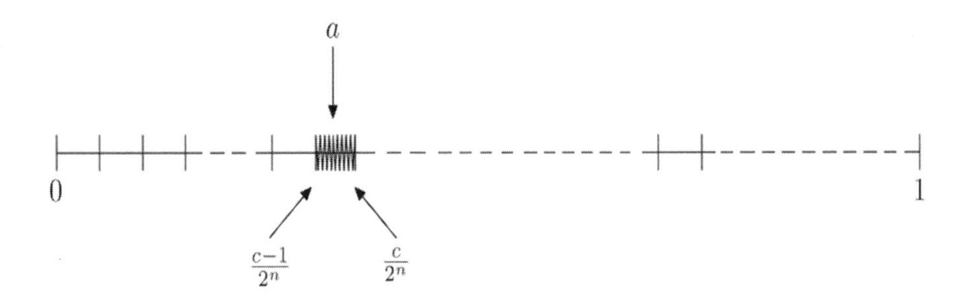

Figure 4.2 a is in the interval J_c

Equivalently, c is the unique integer such that

$$c - 1 \leq 2^n a < c.$$

We are given that $a + P \leq 1$ and $2^n > 1/P$, hence

$$c = (c-1) + 1 < 2^n a + 2^n P = 2^n(a + P) \leq 2^n.$$

Thus $c < 2^n$ and the binary representation of c has the form

$$c = 2^{n-1}z_1 + 2^{n-2}z_2 + \cdots + z_n.$$

Finally, we have shown that $c > 2^n a$ and $c < 2^n(a+P)$, so $c/2^n = 0.z_1z_2\ldots z_n$ lies in the interval $[a, a+P)$ (Figure 4.3.) □

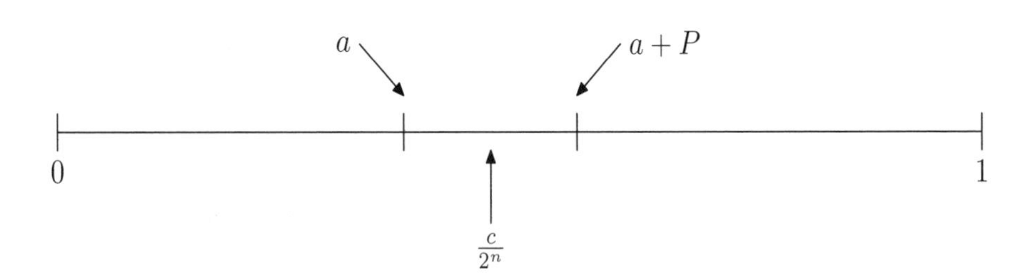

Figure 4.3 $c/2^n$ is in the interval $[a, a+P)$

EXERCISES

4.14. Find the rational numbers corresponding to the words 0111010 and 1001001.

4.15. Explain why the rational number that we normally write as $1/3$ does not correspond to a binary word $z_1z_2\ldots z_n$ for any $n \in \mathbb{N}$.

4.7 Arithmetic coding

We shall use Theorem 4.14 to construct prefix-free binary codes for the strings of symbols that are emitted by a stationary source. In that case there is a probability distribution on the set S^r of strings $X = x_1x_2\cdots x_r$ of length r in S. (In order to avoid cumbersome notation we shall denote the distribution on S^r by P, for any value of r.) In general we do *not* assume the independence property

$$P(x_1x_2\cdots x_r) \textit{ is not necessarily equal to } P(x_1)P(x_2)\cdots P(x_r).$$

We shall suppose that the symbols in the alphabet S are arranged in a fixed order

$$\alpha < \beta < \gamma < \delta < \cdots \quad < \omega.$$

Using this order we can define an order on the strings of symbols, in the same way as the words in a dictionary are ordered using the alphabetical order of the letters.

Definition 4.15 (Dictionary order)

The *dictionary order* on S^r is defined as follows. If X, Y are two different strings in S^r, let i be the least integer such that $x_i \neq y_i$. If $y_i < x_i$ put $Y < X$, otherwise put $X < Y$.

It is convenient to reserve α and ω for the first and last symbols, irrespective of the size of S. For example, if $|S| = 3$, we take then $S = \{\alpha, \beta, \omega\}$, with the order $\alpha < \beta < \omega$. In this case the number of strings of length 4 is $3^4 = 81$. The first seven strings are

$$\alpha\alpha\alpha\alpha < \alpha\alpha\alpha\beta < \alpha\alpha\alpha\omega < \alpha\alpha\beta\alpha < \alpha\alpha\beta\beta < \alpha\alpha\beta\omega < \alpha\alpha\omega\alpha < \ldots$$

and the last seven are

$$\ldots < \omega\omega\alpha\omega < \omega\omega\beta\alpha < \omega\omega\beta\beta < \omega\omega\beta\omega < \omega\omega\omega\alpha < \omega\omega\omega\beta < \omega\omega\omega\omega.$$

Definition 4.16 (Cumulative probability function)

Given a distribution P on S^r we define the associated *cumulative probability function* a on S^r as follows. If X is the first string in S^r put $a(X) = 0$. Otherwise put

$$a(X) = \sum_{Y<X} P(Y).$$

For example, suppose that $S = \{\alpha, \beta, \gamma, \omega\}$ and take $r = 1$. If the values of P on S are

$$P(\alpha) = 0.5, \quad P(\beta) = 0.3, \quad P(\gamma) = 0.1, \quad P(\omega) = 0.1,$$

then the values of a are

$$a(\alpha) = 0.0, \quad a(\beta) = 0.5, \quad a(\gamma) = 0.8, \quad a(\omega) = 0.9.$$

Note that, for each $x \in S$ there is a corresponding interval $[a(x),\ a(x)+P(x))$, and these intervals form a partition of the interval $[0, 1)$ (Figure 4.4).

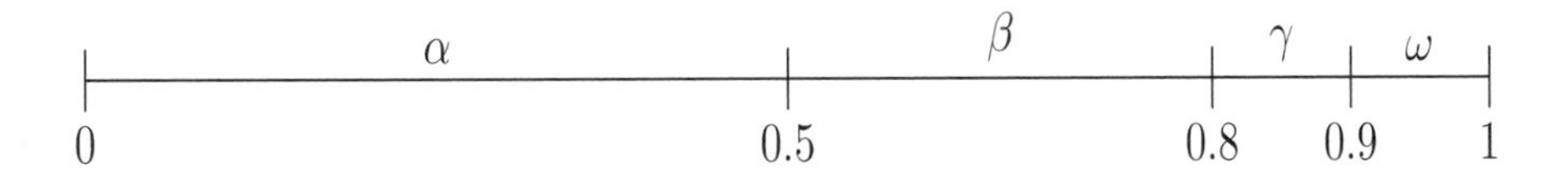

Figure 4.4 The partition of $[0, 1)$ associated with given distribution on S^1

Generally, given a string $X \in S^r$ we denote by I_X the corresponding interval:

$$I_X = [a(X),\ a(X) + P(X)).$$

For each fixed $r \geq 1$ these intervals form a partition of $[0, 1)$.

Definition 4.17 (Arithmetic code)

Suppose that a stationary source emits symbols from a set S and the probability distribution on S^r is known. For $X \in S^r$ let $P = P(X)$, define n_P to be the least integer such that $2^{n_P} \geq 1/P$, and let $n = n_P + 1$. The *arithmetic code* $c : S^r \to \mathbb{B}^*$ is defined by taking $c(X)$ to be the word $z_1 z_2 \cdots z_n$ that represents the unique integer c for which $c - 1 \leq 2^n a(X) < c$.

According to Theorem 4.14, the condition $n = n_P + 1$ ensures that the number $0.z_1 z_2 \ldots z_n$ is in the interval I_X.

Example 4.18

Let $S = \{\alpha, \beta, \gamma, \omega\}$. The following table defines a probability distribution on S^2: for example, the number in row β and column γ is the probability of $\beta\gamma$.

	α	β	γ	ω
α	0.31	0.13	0.04	0.02
β	0.13	0.15	0.01	0.01
γ	0.04	0.01	0.03	0.02
ω	0.02	0.01	0.02	0.05

Determine the arithmetic code for S^2.

Solution For each string $X \in S^2$ we take $P = P(X)$, $a = a(X)$ and calculate $1/P$, n_P, and $n = n_P + 1$. Then we determine c such that $c - 1 \leq 2^n a < c$, and take $c(X)$ to be the binary word of length n corresponding to c. The calculations for the first few codewords are tabulated below. Note that each fraction $c/2^n$ lies in the corresponding interval $[a, a + P)$, as guaranteed by Theorem 4.14.

X	P	a	$1/P$	n_P	n	c	$c(X)$
$\alpha\alpha$	0.31	0.00	3.2	2	3	1	001
$\alpha\beta$	0.13	0.31	7.7	3	4	5	0101
$\alpha\gamma$	0.04	0.44	25	5	6	29	011101
$\alpha\omega$	0.02	0.48	50	6	7	62	0111110
$\beta\alpha$	0.13	0.50	7.7	3	4	9	1001
..	...	...	..	.	.	.	...

It must be stressed that we do not have to calculate the codewords in order. For instance, if we require the codeword for $\gamma\beta$ we can find it directly, knowing only that $P(\gamma\beta) = 0.01$ and $a(\gamma\beta) = 0.84$.

X	P	a	$1/P$	n_P	n	c	$c(X)$
$\gamma\beta$	0.01	0.84	100	7	8	216	11011000

Note that this is not a memoryless source, since (for example)

$$P(\alpha) = P(\alpha\alpha) + P(\alpha\beta) + P(\alpha\gamma) + P(\alpha\omega) = 0.31 + 0.13 + 0.04 + 0.02 = 0.5,$$

and $P(\alpha\alpha) = 0.31 \neq P(\alpha)P(\alpha)$.

EXERCISES

4.16. Find the codeword for $\omega\alpha$ in the arithmetic code for the distribution given in Example 4.18.

4.17. Find the average word-length L_2 for the code discussed in Example 4.18. [You do not need to calculate the codewords.] Compare the values of $L_2/2$, $H(\mathbf{p}^1)$, and $H(\mathbf{p}^2)/2$.

4.18. Construct the arithmetic code for the probability distribution on S^2 given in Example 4.7, taking $a < b < c$. Compare the result with the Huffman code for this source.

4.19. A memoryless source is defined by the distribution P on S, which determines the distributions (also denoted by P) on S^r for all $r \geq 2$. Show that the corresponding cumulative distribution functions satisfy the equation

$$a(x_1x_2\ldots x_r) \;=\; a(x_1x_2\ldots x_{r-1}) + P(x_1x_2\ldots x_{r-1})a(x_r).$$

4.20. Deduce from the previous exercise that, in the memoryless case, the value of $a(X)$ for a string $X \in S^r$ can be computed given only the values of $P(s)$ and $a(s)$ for the symbols s that occur X. Illustrate your answer by writing down a set of equations for calculating $a(\beta\omega\gamma\beta)$ when S contains the symbols $\alpha < \beta < \gamma < \cdots < \omega$.

4.8 The properties of arithmetic coding

We have shown that in arithmetic coding the codewords can be constructed in isolation, and so in practice the process is much more efficient than using Huffman's rule. It remains to establish that arithmetic coding has the other properties that we expect from a workable system. Specifically, we need to show that: (i) the code is prefix-free, so decoding can be done 'on-line', and (ii) the code is close-to-optimal. These facts are consequences of the choice $n = n_P + 1$ in the definition.

Theorem 4.19

An arithmetic code constructed according to Definition 4.17 is prefix-free.

Proof

We must show that if $z_1 z_2 \dots z_n$ is a codeword, say $c(X)$, for $X \in S^r$, then there cannot be another string $Y \in S^r$ such that $c(Y)$ is of the form $z_1 z_2 \dots z_n y_{n+1} \dots y_{n+s}$. We do this by showing that all numbers of the form

$$0.z_1 z_2 \dots z_n y_{n+1} \dots y_{n+s}$$

belong to the same interval I_X as $0.z_1 z_2 \dots z_n$. Thus any such number cannot belong to an interval I_Y for $Y \neq X$, and it cannot represent Y.

By the construction (Theorem 4.14 and Definition 4.17)

$$0.z_1 z_2 \dots z_n \; < \; a(X) + P(X) \; \leq \; a(X) + \frac{1}{2^n}.$$

Thus

$$\begin{aligned}
0.z_1 z_2 \dots z_n y_{n+1} \dots y_{n+s} &= 0.z_1 z_2 \dots z_n + \frac{y_{n+1}}{2^{n+1}} + \cdots + \frac{y_{n+s}}{2^{n+s}} \\
&< a(X) + \frac{1}{2^n} + \frac{1}{2^{n+1}} + \cdots + \frac{1}{2^{n+s}} \\
&< a(X) + \frac{1}{2^n} + \frac{1}{2^n} \\
&= a(X) + \frac{1}{2^{n-1}} \\
&= a(X) + \frac{1}{2^{n_P}} \\
&\leq a(X) + P(X).
\end{aligned}$$

□

Theorem 4.20

Let P be a probability distribution on the set S^r of strings of length r, and let L be the average word-length of the corresponding arithmetic code. Then

$$L < H(P) + 2.$$

Proof

For $X \in S^r$ let n_X be the length of the codeword $c(X)$. As prescribed by the arithmetic coding rule, $n_X - 1$ is the *least* integer such that $2^{n_X - 1} \geq 1/P(X)$. It follows that $2^{n_X - 2} < 1/P(X)$, that is, $n_X - 2 < \log(1/P(X))$. Hence, using the fact that $\sum P(X) = 1$, the average word-length L satisfies

$$L = \sum_X P(X) n_X \; < \; \sum_X P(X)(\log(1/P(X)) + 2) \; = \; H(P) + 2.$$

□

The entropy H for a stationary source is defined to be the greatest lower bound of the numbers $H(P)/r$. Using essentially the same argument as in the proof of Theorem 4.11, it follows that for an arithmetic code the average word-length in *bits per symbol*, L/r, can be made arbitrarily close to H by taking r to be sufficiently large. The fact that an 'extra 2 bits' are allowed by Theorem 4.20 is essentially no worse than the fact that an 'extra 1 bit' may be required by the optimal code.

EXERCISES

4.21. Consider a memoryless source with $S = \{\alpha, \beta, \gamma, \omega\}$ and $P(\alpha) = 0.5$, $P(\beta) = 0.3$, $P(\gamma) = 0.1$, $P(\omega) = 0.1$. Calculate the average word-length of the corresponding arithmetic code for S^2. [You do not need to find the codewords.] Compare this value with the entropy of the source.

4.22. If the source is as in the previous exercise, find the codewords for: (i) $\beta\omega$ in the code for S^2; and (ii) $\alpha\gamma\beta$ in the code for S^3.

4.9 Coding with a dynamic dictionary

Definition 4.21 (Dictionary, index)

A *dictionary* D based on the alphabet S is a sequence of distinct words in S^*:

$$D = d_1, d_2, d_3, \ \ldots \ , d_N,$$

We say that the *index* of d_i is i.

In arithmetic coding we imposed an 'alphabetical order' on the set of symbols S, and constructed the corresponding 'dictionary order' on S^r. For example, if S has three symbols in the order $\alpha < \beta < \omega$, the set S^2 is a dictionary with the order

$$\alpha\alpha, \alpha\beta, \alpha\omega, \beta\alpha, \beta\beta, \beta\omega, \omega\alpha, \omega\beta, \omega\omega,$$

and the index of the element $\omega\alpha$ is 7. This is an example of a *static* dictionary.

In this section we consider a method of coding in which the dictionary is *dynamic* – in other words it is constructed so that it contains those strings that do occur in a given message. It is assumed that the source is stationary so that strings of symbols which occur near the beginning of the message are typical of those that will occur later on, but no specific information about the distribution of the symbols is needed.

We shall describe the Lempel-Ziv-Welch coding system, also known as LZW. The alphabet S has size m, with symbols $s_1 < s_2 < \cdots < s_m$. Both the encoder and the decoder begin with the dictionary

$$D_0 = d_1, d_2, \ldots, d_m, \qquad \text{where } d_i = s_i \ (i = 1, 2, \ldots, m).$$

If the message is $X = x_1 x_2 \cdots x_n$, where $x_i \in S$ $(i = 1, 2, \ldots, n)$, the encoder makes a coded form of the message, $c(X)$, and at the same time constructs a dictionary D_X by adding to D_0 certain strings $s_p s_q \ldots$ that occur in X. The encoding rule replaces each such string by its index in D_X. The decoder uses the sequence of numbers $c(X)$, and the initial dictionary D_0, to reconstruct X and (incidentally) D_X.

Definition 4.22 (LZW encoding)

Suppose that a message $X = x_1 x_2 \cdots x_n$ in the alphabet $S = \{s_1, s_2, \ldots, s_m\}$ is given. Let $D_0 = d_1, d_2, \ldots, d_m$, where $d_i = s_i$ $(i = 1, 2, \ldots, m)$. The *LZW encoding* rules construct $c(X) = c_1\, c_2\, c_3\, \ldots$ in a series of steps.

- Step 1 The first symbol x_1 $(= s_p$ say) is an entry d_p in D_0. Encode x_1 by defining
$$c_1 = p.$$
The string $x_1 x_2$ $(= s_p s_q$ say) is not in D_0. Define
$$d_{m+1} = x_1 x_2, \qquad D_1 = (D_0, d_{m+1}).$$

- Step k $(k \geq 2)$ Suppose that Steps $1, 2, \ldots, k-1$ have been completed. This means that the code $c_1\, c_2\, \ldots, c_{k-1}$ for an initial segment $x_1 x_2 \cdots x_i$ of X, and a dictionary $D_{k-1} = d_1, d_2, \ldots, d_{m+k-1}$, have been constructed. Find the longest string w of the form
$$w = x_{i+1} \cdots x_j \quad (j \geq i+1)$$
such that w is in D_{k-1}, say $w = d_t$. Such a string certainly exists, because x_{i+1} is a single symbol and it is already in the initial dictionary D_0. By definition, wx_{j+1} is not in D_{k-1}. Encode the segment $x_{i+1} \cdots x_j$ of X by
$$c_k = t, \quad \text{and define } d_{m+k} = wx_{j+1}, \quad D_k = (D_{k-1}, d_{m+k}).$$

- Repeat the procedure until the end of the message is reached.

Example 4.23

Take the alphabet to be $S = \{\mathtt{A}, \mathtt{B}, \mathtt{C}, \mathtt{D}, \mathtt{R}\}$ and apply the LZW encoding rules to the message

$$X = \mathtt{ABRACADABRA} \quad .$$

Solution We start with the dictionary $D_0 = \mathtt{A}, \mathtt{B}, \mathtt{C}, \mathtt{D}, \mathtt{R}$. The rule for Step 1 tells us to encode A, which has index 1, and add AB to the dictionary:

$$c(\mathtt{A}) = 1, \qquad D_1 = \mathtt{A}, \mathtt{B}, \mathtt{C}, \mathtt{D}, \mathtt{R}, \mathtt{AB}.$$

At Step 2, B is in D_1 but BR is not, so we encode B by its index 2, and add BR:

$$c(\mathtt{AB}) = 1\ 2, \qquad D_2 = \mathtt{A}, \mathtt{B}, \mathtt{C}, \mathtt{D}, \mathtt{R}, \mathtt{AB}, \mathtt{BR}.$$

The rules say that at Steps 3,4,5,6,7 we continue in a similar way until we reach the stage:

$$c(\mathtt{ABRACAD}) = 1\ 2\ 5\ 1\ 3\ 1\ 4, \qquad D_7 = \mathtt{A}, \mathtt{B}, \mathtt{C}, \mathtt{D}, \mathtt{R}, \mathtt{AB}, \mathtt{BR}, \mathtt{RA}, \mathtt{AC}, \mathtt{CA}, \mathtt{AD}, \mathtt{DA}.$$

At Step 8 AB is already in D_7, with index 6, but ABR is not in D_7. So the outcome is

$$c(\mathtt{ABRACADAB}) = 1\ 2\ 5\ 1\ 3\ 1\ 4\ 6,$$

$$D_8 = \mathtt{A}, \mathtt{B}, \mathtt{C}, \mathtt{D}, \mathtt{R}, \mathtt{AB}, \mathtt{BR}, \mathtt{RA}, \mathtt{AC}, \mathtt{CA}, \mathtt{AD}, \mathtt{DA}, \mathtt{ABR}.$$

At Step 9, RA is in the dictionary, with index 8, and the message ends, so

$$c(\mathtt{ABRACADABRA}) = 1\ 2\ 5\ 1\ 3\ 1\ 4\ 6\ 8.$$

At first sight it is not obvious that an LZW code is uniquely decodable. The decoder must emulate the encoder by adding a new entry to the dictionary at each step, and always seems to be one step behind. This problem can occur at Step 1, and it is worth looking at this case carefully.

Suppose the decoder is given D_0 and the encoded message

$$C = c_1\, c_2 \cdots .$$

Step 1 in the LZW encoding rules says that $c_1 = p$, where $d_p = s_p$ is in D_0, so c_1 is decoded by setting $x_1 = s_p$. In order to complete Step 1, the decoder must also decide how to augment D_0 by adding a new string d_{m+1}, and this requires consideration of c_2.

Let $c_2 = r$. If $r \leq m$ the procedure is simple, because d_r is in D_0, and $d_r = s_r$. The message must therefore begin with $x_1 x_2 = s_p s_r$, and the decoder emulates the encoder by adding this string to the dictionary. But what if $r > m$? In this 'awkward case' r can only be $m+1$, because c_2 is constructed using the dictionary D_1. According to the rule for Step 1, $d_{m+1} = s_p s_q$ where $x_1 = s_p$, $x_2 = s_q$. Now the decoder can identify x_2, because $c_1\, c_2$ is the encoded form of $s_p s_p s_q$. Thus $x_2 = s_p$ so $q = p$ and the new entry in the dictionary is actually $d_{m+1} = s_p s_p$.

A similar argument works at every step, as in the proof of the following theorem.

Theorem 4.24

An LZW code constructed as in Definition 4.22 is uniquely decodable.

Proof

Suppose that the partial code $c_1\, c_2\ \ldots\ c_{k-1}$ has been successfully decoded as $x_1 x_2 \ldots x_i$. Suppose also that the dictionary D_{k-1} has been constructed by adding $k-1$ strings to D_0. The argument given above shows that these assumptions are justified when $k-1=1$, so we may suppose that $k \geq 2$. We must give the rules for decoding c_k and constructing $D_k = (D_{k-1}, d_{m+k})$.

The encoding rules imply that c_k is the index of a string $s_u \ldots s_v$ in D_{k-1}. Thus c_k is decoded by putting $x_{i+1} = s_u, \ldots, x_j = s_v$.

In order to construct the new dictionary entry d_{m+k}, the encoder must use the value of c_{k+1}. If $c_{k+1} = r \leq m+k-1$ then d_r is in D_{k-1}, say $d_r = s_a \ldots$. Thus $x_{i+1} = s_a$, and the new dictionary entry is $d_{m+k} = s_u \ldots s_v s_a$. Since the construction of c_{k+1} uses only D_k, the only other possibility is the 'awkward case' $c_{k+1} = m + k$. Now d_{m+k} is a string of the form $s_u \ldots s_v s_z$, where $s_z = x_{j+1}$. So in this case $c_k\, c_{k+1}$ is the encoded form of $s_u \ldots s_v s_u \ldots s_v s_z$, and in fact $x_{j+1} = s_u$. In other words $d_{m+k} = s_u \ldots s_v s_u$. This completes the emulation of Step k, and the proof is complete.

□

Example 4.25

Given the dictionary $D_0 = \mathtt{I,M,P,S}$, decode the message

$$2\ 1\ 4\ 4\ 6\ 8\ 3\ 3\ 1 \quad .$$

Solution At Step 1, $c_1 = 2$ and $d_2 = \mathtt{M}$, so $x_1 = \mathtt{M}$. Also $c_2 = 1$ and $d_1 = \mathtt{I}$, so the new dictionary entry is MI. Thus

$$2 \mapsto \mathtt{M}, \qquad D_1 = \mathtt{I,M,P,S,MI}.$$

At Step 2, $c_2 = 1$ and $d_1 = \texttt{M}$, so $x_2 = \texttt{I}$. Also $c_3 = 4$ and $d_4 = \texttt{S}$, so the new dictionary entry is IS. Thus

$$2\,1 \mapsto \texttt{MI}, \qquad D_2 = \texttt{I},\texttt{M},\texttt{P},\texttt{S},\texttt{MI},\texttt{IS}.$$

Steps 3 and 4 are similar, resulting in

$$2\,1\,4\,4 \mapsto \texttt{MISS}, \qquad D_4 = \texttt{I},\texttt{M},\texttt{P},\texttt{S},\texttt{MI},\texttt{IS},\texttt{SS},\texttt{SI}.$$

At the next step $c_5 = 6$, and $d_6 = \texttt{IS}$ so $x_5 = \texttt{I}, x_6 = \texttt{S}$. Also $c_6 = 8$ and $d_8 = \texttt{SI}$ so the new dictionary entry is ISS. Thus

$$2\,1\,4\,4\,6 \mapsto \texttt{MISSIS}, \qquad D_5 = \texttt{I},\texttt{M},\texttt{P},\texttt{S},\texttt{MI},\texttt{IS},\texttt{SS},\texttt{SI},\texttt{ISS}.$$

Steps 7,8,9 are similar, and the result is

$$2\,1\,4\,4\,6\,8\,3\,3\,1 \mapsto \texttt{MISSISSIPPI}.$$

At this point, the reader is probably asking whether the LZW rules actually achieve significant compression. In the toy examples given above, only a very small amount of compression is achieved, at some cost. For example, the message ABRACADABRA has length 11, and the coded form 1 2 5 1 3 1 4 6 8 has length 9, but the reduction is achieved by using a nontrivial calculation. However the examples do suggest that, as the message increases in length, with inevitably more repetition of strings, then there may be some improvement.

In practice, LZW coding works well, and its widespread use confirms that the principle is sound. The presentation of the rules given above is intended to make them easily understood, but it is possible to streamline the calculations by using more efficient data structures.

EXERCISES

4.23. Verify that the LZW encoding rules for MISSISSIPPI with the initial dictionary I,M,P,S produce the code 2 1 4 4 6 8 3 3 1.

4.24. Verify that the LZW decoding rules for 1 2 5 1 3 1 4 6 8 with the initial dictionary A,B,C,D,R produce the message ABRACADABRA.

4.25. Suppose the initial dictionary is B,D,E,N,O,R,T,$\sqcup$. Decode the following message.

1 3 7 8 5 4 8 7 3 14 5 6 8 9 11 13 12 4 3 12 20 2 25 5 11 22

4.26. Suppose we are given $D_0 = a, b, c$ and the coded message 1 2 1 3 4 7 9. Show that at Step $k = 6$ of the decoding procedure we encounter the 'awkward case' in which the codeword c_{k+1} is not the index of an entry in the dictionary D_{k-1}. Explain how the difficulty is resolved.

Further reading for Chapter 4

The basic ideas of arithmetic coding were known to the founding fathers of information theory, but it was not until the advances in computer technology in the 1970s that there was serious interest in data compression techniques. The influential paper by Gallager [**4.1**] discussed various possible methods of adaptive coding. Slightly earlier, Jacob Ziv and Abraham Lempel [**4.4**] had suggested the idea of coding with a dictionary. There are several variants of their idea; the one known as LZW that is discussed in this book was developed by Welch in 1984 [**4.3**]. The LZW algorithm is now used in many applications, including `compress`, `gzip`, and Acrobat. A standard reference for the practical side of data compression is the book by Sayood [**4.2**], and MacKay [**3.3**, Chapter 6] gives an interesting commentary on the respective merits of the various approaches.

In this book we discuss stationary sources because they provide the simplest model that occupies the common ground between theory and practice. There are more sophisticated models, such as a *Markov source* and an *ergodic source*, and these too have been the subject of serious study by mathematicians. An excellent introduction is given by Welsh [**3.5**].

It is traditional to define the entropy H of a stationary source to be the *limit* of the sequence $H(\mathbf{p}^r)/r$, rather than the infimum. This creates a problem, because it is not obvious that the limit exists, whereas the existence of the infimum is a basic property of the real number system. In fact the two quantities are equal, but the proof is rather complicated (see Exercise 4.11). All the important properties of H can be proved using the simpler definition.

4.1 R.G. Gallager. Variations on a theme by Huffman. *IEEE Trans. Info. Theory* IT-24 (1978) 668-674.

4.2 N. Sayood. *Introduction to Data Compression*, Third edition. Morgan-Kaufmann (2005).

4.3 T.A. Welch. A technique for high performance data compression. *IEEE Computer* 17 (1984) 8-19.

4.4 J. Ziv and A. Lempel. A universal algorithm for sequential data compression. *IEEE Trans. Info. Theory* IT-23 (1977) 337-343.

5
Noisy channels

5.1 The definition of a channel

Let I be a finite alphabet. We consider the symbols in I as inputs to a device, referred to as a *channel*, which transmits them in such a way that errors may occur. As a result of these errors, which we call *noise*, the symbols that form the output of the device belong to another set J. The sets I and J may be the same, but that does not mean that a specific input $i \in I$ will result in the output of the same symbol. This situation is illustrated in Figure 5.1.

Figure 5.1 A diagrammatic representation of a 'channel'

For each $i \in I$ and $j \in J$, we shall by denote by $\Pr(j \mid i)$ the conditional probability that the output is j, given that the input is i:

$$\Pr(j \mid i) \;=\; \Pr(\text{output is } j \mid \text{input is } i).$$

N. L. Biggs, *An Introduction to Information Communication and Cryptography*,
DOI: 10.1007/978-1-84800-273-9_5, © Springer-Verlag London Limited 2008

Definition 5.1 (Noisy channel)

A *channel* Γ with input set I and output set J is a matrix whose entries are the numbers $\Pr(j \mid i)$:

$$\Gamma_{ij} = \Pr(j \mid i) \qquad (i \in I, j \in J).$$

(Note that the entries of Γ are labelled so that the rows correspond to inputs and the columns correspond to outputs.) If at least one of the terms Γ_{ij} with $i \neq j$ is non-zero, we say that the channel is *noisy*.

From a mathematical viewpoint, there is no distinction between a *channel* and the corresponding *channel matrix*, and we shall use these terms interchangeably.

Definition 5.2 (Binary Symmetric Channel)

A *Binary Symmetric Channel (BSC)* corresponds to a matrix of the form

$$\Gamma = \begin{pmatrix} \Gamma_{00} & \Gamma_{01} \\ \Gamma_{10} & \Gamma_{11} \end{pmatrix} = \begin{pmatrix} 1-e & e \\ e & 1-e \end{pmatrix}.$$

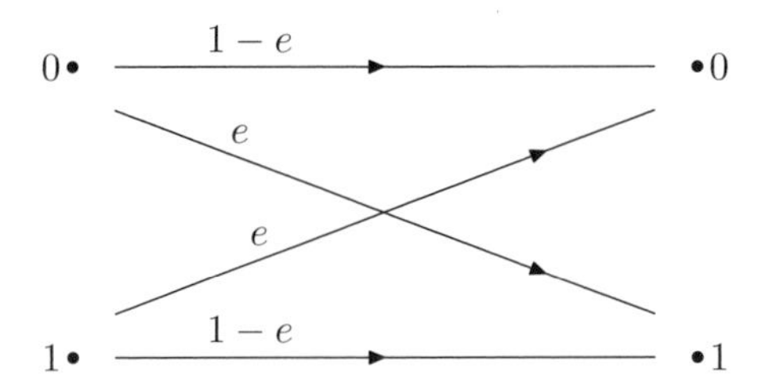

Figure 5.2 The binary symmetric channel with bit-error probability e

A Binary Symmetric Channel is illustrated in Figure 5.2. The rows and columns of the channel matrix Γ correspond to 0 and 1, the elements of the binary alphabet $\mathbb{B}$. The fact that $\Gamma_{01} = \Gamma_{10} = e$ means that the output symbol differs from the input symbol with probability e, and we refer to e as the *bit-error probability*. The probability that a symbol is transmitted correctly is $\Gamma_{00} = \Gamma_{11} = 1 - e$. Usually we assume that e is a small positive number: for example, $e = 0.01$ means that one bit in every hundred is transmitted wrongly. Of course, that would be unacceptably *large* in most real situations.

Here is a different kind of channel.

Example 5.3

Figure 5.3 illustrates a simplified keypad with six keys A, B, C, D, E, F. Let us say that two keys are adjacent if an edge of one is next to an edge of the other. Given any two adjacent keys x and y there is a probability 0.1 that when I intend to enter x I shall press y. Write down the channel matrix for this situation. If I intend to enter $FACE$, what is the probability that the keypad will register correctly?

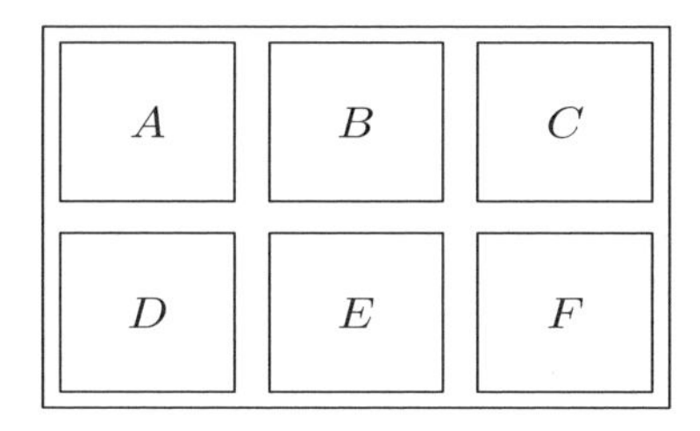

Figure 5.3 A simplified keypad

Solution The input and output sets are $\{A, B, C, D, E, F\}$, and the channel matrix is

$$\Gamma = \begin{pmatrix} 0.8 & 0.1 & 0 & 0.1 & 0 & 0 \\ 0.1 & 0.7 & 0.1 & 0 & 0.1 & 0 \\ 0 & 0.1 & 0.8 & 0 & 0 & 0.1 \\ 0.1 & 0 & 0.0 & 0.8 & 0.1 & 0 \\ 0 & 0.1 & 0 & 0.1 & 0.7 & 0.1 \\ 0 & 0 & 0.1 & 0 & 0.1 & 0.8 \end{pmatrix}.$$

The probability of correctly registering $FACE$ is $0.8 \times 0.8 \times 0.8 \times 0.7 = 0.3584$.

Lemma 5.4

Each row of a channel matrix has sum equal to 1. That is,

$$\sum_{j \in J} \Gamma_{ij} = 1 \quad \text{for all } i \in I.$$

Proof

By definition,

$$\sum_{j \in J} \Gamma_{ij} = \sum_{j \in J} \Pr(j \mid i).$$

The right-hand side represents the total probability of all outputs, given that the input is i, and so it is equal to 1. □

EXERCISES

5.1. A *binary asymmetric channel* is similar to a BSC, except that the probability of error when 0 is sent is a, and the probability of error when 1 is sent is b, where $a \neq b$. Write down the matrix for this channel.

5.2. A *ternary symmetric channel* is a channel for which the input set I and output set J are both $\{0, 1, 2\}$, and the probability that a symbol i becomes $j \neq i$ is x. Write down the matrix for this channel.

5.2 Transmitting a source through a channel

Suppose that a symbol i in the input set I occurs with probability p_i, so that we have a probability distribution $\mathbf{p}$ on I. Then we can think of the input to the channel Γ as being generated by a source $(I, \mathbf{p})$. Similarly, letting q_j be the probability that the output symbol is j, the output can be regarded as a source $(J, \mathbf{q})$. The following theorem describes the relationship between $\mathbf{p}$ and $\mathbf{q}$.

Theorem 5.5

Let Γ be a channel matrix, and let the distributions associated with the input source $(I, \mathbf{p})$ and the output source $(J, \mathbf{q})$ be written as row vectors,

$$\mathbf{p} = [p_1, p_2, \ldots, p_m], \qquad \mathbf{q} = [q_1, q_2, \ldots, q_n].$$

Then

$$\mathbf{q} = \mathbf{p}\Gamma.$$

Proof

For each $i \in I$ and $j \in J$ let t_{ij} denote the probability of the event that the input to Γ is i and the output is j. According to the addition law, it follows that

$$q_j = \sum_{i \in I} t_{ij}.$$

Also, by the definition of conditional probability,

$$t_{ij} = \Pr(\text{output is } j \mid \text{input is } i) \times \Pr(\text{input is } i) = \Gamma_{ij} p_i.$$

Thus

$$q_j = \sum_{i \in I} p_i \Gamma_{ij} = (\mathbf{p}\Gamma)_j,$$

that is,

$$\mathbf{q} = \mathbf{p}\Gamma.$$

□

We now have a model in which the channel Γ can be regarded as a link between two sources, $(I, \mathbf{p})$ and $(J, \mathbf{q})$. The first source is produced by a *Sender*, while the second source, the result of transmission through Γ, is available to a *Receiver*. These sources are related by the equation $\mathbf{q} = \mathbf{p}\Gamma$.

Example 5.6

Write down explicitly the equations linking $\mathbf{p} = [p_0, p_1]$ and $\mathbf{q} = [q_0, q_1]$ when Γ is the BSC with bit-error probability e. If $e = 0.1$ and $\mathbf{p} = [0.7, 0.3]$, what is $\mathbf{q}$?

Solution The matrix equation is

$$\begin{pmatrix} q_0 & q_1 \end{pmatrix} = \begin{pmatrix} p_0 & p_1 \end{pmatrix} \begin{pmatrix} 1-e & e \\ e & 1-e \end{pmatrix},$$

which is equivalent to the equations

$$\begin{aligned} q_0 &= p_0(1-e) + p_1 e \\ q_1 &= p_0 e + p_1(1-e). \end{aligned}$$

If $e = 0.1$ and $\mathbf{p} = [0.7, 0.3]$ then $\mathbf{q} = [0.66, 0.34]$.

EXERCISES

5.3. Suppose the output symbols from a channel Γ_1 are the input symbols for channel Γ_2, and Γ is the resulting combined channel. In this situation we say that Γ is the result of combining Γ_1 and Γ_2 *in series*. What is the relationship between the matrices for Γ_1, Γ_2, and Γ?

5.4. Suppose that a BSC with bit-error probability e and a BSC with bit-error probability e' are combined in series. Show that the result is a BSC, and find its bit-error probability.

5.5. Suppose that n binary symmetric channels with the same bit-error probability e are combined in series. If $0 < e < 1$, what is the 'limit' channel as $n \to \infty$? What happens if $e = 0$? What happens if $e = 1$?

5.6. Consider the binary symmetric channel with bit-error probability $e = 0.01$. If the input has probability distribution $\mathbf{p} = [0.6, 0.4]$ what is the output distribution $\mathbf{q}$? Compare the entropies of the input and output.

5.7. Consider a binary asymmetric channel (Exercise 5.1) with $a = 0.02$, $b = 0.04$. Write down explicitly the equations for $\mathbf{q}$ in terms of $\mathbf{p}$.

5.8. It is observed that in the output from a binary asymmetric channel with $a \neq b$ the symbols 0 and 1 occur equally often. Find the probability distribution on the input, and show that the entropy of the output exceeds that of the input.

5.3 Conditional entropy

The existence of errors tends to equalize probabilities, because symbols that occur more frequently are transmitted wrongly more often. In Example 5.6 the uncertainty of the input and the output are, respectively,

$$h(0.7) \approx 0.881, \qquad h(0.66) \approx 0.925,$$

where h is the standard entropy function. As we might expect, uncertainty is increased by transmission through a noisy channel. More precisely, from the viewpoint of an observer who has access to both input and output, the effect of transmitting a source through a noisy channel Γ is to increase the uncertainty.

A more subtle problem is to describe the situation from the viewpoint of the Receiver, for whom the output provides the only available information about the input.

Question: How much information about the input is available to a Receiver who knows the output of Γ?

In order to answer this question, it is helpful to reconsider the situation. We set up the model in terms of an *input* that is passed through a *channel* to produce an *output*, because that is the way an engineer would think of it. But from a mathematical viewpoint the fundamental object is simply a probability distribution $\mathbf{t}$ on the set $I \times J$:

$$t_{ij} = \Pr(\text{input is } i \text{ and output is } j).$$

We have already used $\mathbf{t}$ in the proof of Theorem 5.5. The significant point is that, when $\mathbf{t}$ is given, all the other quantities can be derived from it, by using the equations

$$p_i = \sum_j t_{ij}, \qquad q_j = \sum_i t_{ij}, \qquad \Gamma_{ij} = t_{ij}/p_i.$$

In particular, $\mathbf{p}$ and $\mathbf{q}$ are the marginal distributions associated with $\mathbf{t}$.

Definition 5.7 (Conditional entropy)

With the notations as above, the *conditional entropy* $H(\mathbf{p} \mid \mathbf{q})$ is defined to be

$$H(\mathbf{p} \mid \mathbf{q}) \;=\; H(\mathbf{t}) - H(\mathbf{q}).$$

The motivation is that $H(\mathbf{t})$ measures the uncertainty about the input-output pair (i, j), and $H(\mathbf{q})$ measures the uncertainty about the output j. Thus, subtracting the second quantity from the first represents the uncertainty of a Receiver who knows the output and is trying to determine the input. It is worth noting that an alternative definition, more complicated but equivalent, is often used for $H(\mathbf{p} \mid \mathbf{q})$ – see Section 5.5, particularly Exercise 5.18.

Since the input and output distributions are linked by the equation $\mathbf{q} = \mathbf{p}\Gamma$, it follows that $H(\mathbf{p} \mid \mathbf{q})$ depends only on Γ and $\mathbf{p}$. We shall often refer to this quantity as the *conditional entropy of* $\mathbf{p}$ *with respect to transmission through* Γ, and denote it by $H(\Gamma; \mathbf{p})$:

$$H(\Gamma; \mathbf{p}) \;=\; H(\mathbf{p} \mid \mathbf{q}) \qquad (\text{where } \mathbf{q} = \mathbf{p}\Gamma).$$

Example 5.8

Let Γ be the BSC with bit-error probability $e = 0.1$, and let the source distribution be $\mathbf{p} = [0.7, 0.3]$. Calculate $H(\Gamma; \mathbf{p})$.

Solution Since $t_{ij} = p_i \Gamma_{ij}$ we have

$$t_{00} = 0.63, \quad t_{01} = 0.07, \quad t_{10} = 0.03, \quad t_{11} = 0.27.$$

It follows that $H(\mathbf{t}) \approx 1.350$. Since $q_j = \sum_i t_{ij}$ we have $\mathbf{q} = [0.66, 0.34]$, and $H(\mathbf{q}) \approx 0.925$. Hence

$$H(\Gamma; \mathbf{p}) = H(\mathbf{t}) - H(\mathbf{q}) \approx 1.350 - 0.925 = 0.425.$$

The general result for a binary symmetric channel is as follows.

Theorem 5.9

Let Γ be the BSC with bit-error probability e, and let $\mathbf{p}$ be the source distribution $[p_0, p_1] = [p, 1-p]$. Then

$$H(\Gamma; \mathbf{p}) \ = \ h(p) + h(e) - h(q),$$

where $q = p(1-e) + (1-p)e$, and h is the standard entropy function defined by $h(x) = x \log_2(1/x) + (1-x)\log_2(1/(1-x))$.

Proof

The values of $t_{ij} = p_i \Gamma_{ij}$ are

$$t_{00} = p(1-e), \quad t_{01} = pe, \quad t_{10} = (1-p)e, \quad t_{11} = (1-p)(1-e).$$

These values mean that the probability distribution $\mathbf{t}$ on $\mathbb{B}^2$ is simply the product of independent distributions $[p, 1-p]$ and $[1-e, e]$. Hence, by Theorem 4.4, $H(\mathbf{t})$ is the sum of the entropies of these distributions, and

$$H(\Gamma; \mathbf{p}) = H(\mathbf{t}) - H(\mathbf{q}) = h(p) + h(e) - h(q).$$

□

EXERCISES

5.9. Consider the binary symmetric channel with bit-error probability $e = 0.01$ and input distribution $\mathbf{p} = [0.6, 0.4]$ (Exercise 5.6). Calculate the joint distribution $\mathbf{t}$ and hence find the conditional entropy $H(\Gamma; \mathbf{p})$.

5.10. Verify that your answer to the previous exercise agrees with the formula in Theorem 5.9.

5.11. The *binary erasure channel* accepts input symbols 0, 1. The output symbol is the same as the input symbol with probability c and a query (?) with probability $1-c$. In other words, the channel matrix is

$$\Gamma = \begin{pmatrix} c & 0 & 1-c \\ 0 & c & 1-c \end{pmatrix}.$$

If the input distribution is $\mathbf{p} = [p, 1-p]$, show that

$$H(\Gamma; \mathbf{p}) = (1 - c)h(p).$$

5.12. Consider a general binary channel with $I = J = \mathbb{B}$. If the joint distribution on $I \times J$ is given by

$$t_{00} = a, \quad t_{01} = b, \quad t_{10} = c, \quad t_{11} = d, \qquad a + b + c + d = 1,$$

write down $\mathbf{p}$, $\mathbf{q}$, and Γ, and verify that $\mathbf{q} = \mathbf{p}\Gamma$.

5.4 The capacity of a channel

It is fairly obvious that an oil pipeline or a road has a 'capacity', beyond which it cannot function effectively. But in the case of a communication channel this feature is not so obvious. In fact, a good definition does exist, and significant results can be proved about it.

We have defined

- $H(\mathbf{p})$: the uncertainty about symbols emitted by the input source $\mathbf{p}$;
- $H(\Gamma; \mathbf{p})$: the uncertainty about symbols emitted by $\mathbf{p}$ from the viewpoint of a Receiver who knows the symbols emitted by the output source $\mathbf{q} = \mathbf{p}\Gamma$.

The following theorem confirms that these quantities are related in the way that we should expect.

Theorem 5.10

Let Γ be a channel and $\mathbf{p}$ an input distribution for Γ. Then

$$H(\Gamma; \mathbf{p}) \leq H(\mathbf{p}).$$

Equality holds if and only if $\mathbf{p}$ and $\mathbf{q} = \mathbf{p}\Gamma$ are independent distributions.

Proof

The distributions $\mathbf{p}$ and $\mathbf{q}$ are the marginal distributions associated with the joint distribution $\mathbf{t}$. It follows from Theorem 4.4 that $H(\mathbf{t}) \leq H(\mathbf{p}) + H(\mathbf{q})$, with equality if and only if $\mathbf{p}$ and $\mathbf{q}$ are independent. By definition, $H(\Gamma; \mathbf{p}) = H(\mathbf{t}) - H(\mathbf{q})$, and so

$$H(\Gamma; \mathbf{p}) + H(\mathbf{q}) = H(\mathbf{t}) \leq H(\mathbf{p}) + H(\mathbf{q}),$$

with equality if and only if $\mathbf{p}$ and $\mathbf{q}$ are independent. Hence the result. □

Let

$$f_\Gamma(\mathbf{p}) \;=\; H(\mathbf{p}) - H(\Gamma;\mathbf{p}).$$

Since $f_\Gamma(\mathbf{p})$ is a difference between two measures of uncertainty, we can think of it as a measure of *information*. Specifically, it represents the information about symbols emitted by the input source $\mathbf{p}$ that is available to the Receiver who knows the symbols emitted by the output source. For example, suppose Γ is the BSC with bit-error probability $e = 0.1$, and the input distribution is $\mathbf{p} = [0.7, 0.3]$. In Section 5.3 we found that $H(\mathbf{p}) = h(0.7) \approx 0.881$ and $H(\Gamma;\mathbf{p}) \approx 0.425$, so

$$f_\Gamma(\mathbf{p}) \;=\; H(\mathbf{p}) - H(\Gamma;\mathbf{p}) \approx 0.881 - 0.425 = 0.456.$$

Suppose we are given a channel Γ that accepts an input alphabet I of size m; in other words, the channel matrix has m rows. Then for each probability distribution $\mathbf{p}$ on I we have a value of $f_\Gamma(\mathbf{p})$.

Definition 5.11 (Capacity)

The *capacity* γ of Γ is the maximum value of $f_\Gamma(\mathbf{p})$, taken over the set $\mathcal{P}$ of all probability distributions on a set of size m (the number of rows of Γ). That is,

$$\gamma = \max_{\mathcal{P}} f_\Gamma(\mathbf{p}) = \max_{\mathcal{P}} \Big(H(\mathbf{p}) - H(\Gamma;\mathbf{p})\Big).$$

It is important to stress that the maximum exists. Technically this is because f_Γ is a continuous function, and the set of all probability distributions,

$$\mathcal{P} \;=\; \{\mathbf{p} \in \mathbb{R}^m \mid p_1 + p_2 + \cdots + p_m = 1,\; 0 \le p_i \le 1\},$$

is a compact set. For example, when $m = 3$ the set $\mathcal{P}$ is the subset of $\mathbb{R}^3$ indicated by the shaded area in Figure 5.4.

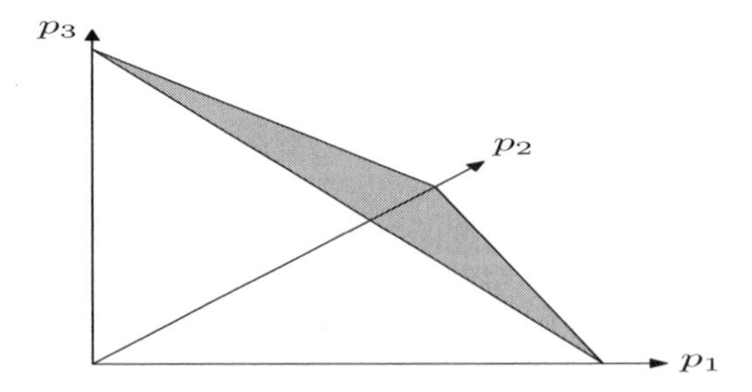

Figure 5.4 The set $\mathcal{P}$ when $m = 3$

The definition of capacity is fundamental. Its full significance will appear in the following chapters, when we discuss some famous results about the transmission of coded messages through noisy channels.

EXERCISES

5.13. Find the maximum values of the following functions on the set $\mathcal{P}$ when $m = 3$.

$$\text{(i) } p_1p_2p_3; \qquad \text{(ii) } p_1^2 + p_2^2 + p_3^2; \qquad \text{(iii) } p_1 + 2p_2 + 3p_3.$$

5.14. If $\mathbf{p} = [p_1, p_2, \ldots, p_m]$, find the maximum value of $H(\mathbf{p})$ on the set $\mathcal{P}$. [Calculus methods are not required.]

5.5 Calculating the capacity of a channel

In general, calculating the capacity of a channel is difficult, but for the BSC there is a simple answer.

Theorem 5.12

The capacity of the BSC with bit-error probability e $(0 \leq e < \frac{1}{2})$ is

$$\gamma = 1 - h(e),$$

where $h(e) = e\log_2(1/e) + (1-e)\log_2(1/(1-e))$.

Proof

Writing $\mathbf{p} = [p, 1-p]$ we have $H(\mathbf{p}) = h(p)$, and according to Theorem 5.9

$$H(\Gamma;\mathbf{p}) = h(p) + h(e) - h(q),$$

where $q = p(1-e) + (1-p)e$. Thus the capacity is the maximum of

$$h(p) - \Big(h(p) + h(e) - h(q)\Big) = h(q) - h(e)$$

taken over the range $0 \leq p \leq 1$.

Since e is a given constant, the maximum occurs when the first term $h(q)$ is a maximum. This occurs when $q = \frac{1}{2}$, that is

$$q = p(1-e) + (1-p)e = \frac{1}{2}.$$

Rearranging this equation we get

$$(p - \frac{1}{2})(1-2e) = 0.$$

So, for any $e < \frac{1}{2}$, the maximum of $h(q)$ occurs when $p = \frac{1}{2}$, and the maximum value is $h(\frac{1}{2}) = 1$. The capacity is therefore

$$\gamma = \max\Big(h(q) - h(e)\Big) = 1 - h(e).$$

□

In most real situations, e is small, so $h(e)$ is close to 0, and γ is close to 1. For example, the capacity of the BSC with bit-error probability 0.01 is

$$\begin{aligned}\gamma = 1 - h(0.01) &= 1 - 0.01\log_2(1/0.01) - 0.99\log_2(1/0.99)\\ &= 0.919.\end{aligned}$$

The definition of channel capacity (Definition 5.11) is formulated in terms of finding the maximum of

$$H(\mathbf{p}) - H(\varGamma;\mathbf{p}), \qquad \text{that is} \qquad H(\mathbf{p}) - H(\mathbf{p} \mid \mathbf{q}).$$

We chose this formulation because it motivates the interpretation of capacity as the maximum amount of information that can be transmitted. However, there are other formulations that are mathematically equivalent, and sometimes more convenient for the purposes of calculation.

Since $H(\mathbf{p} \mid \mathbf{q}) = H(\mathbf{t}) - H(\mathbf{q})$, the quantity to be maximized can be expressed more symmetrically as

$$H(\mathbf{p}) + H(\mathbf{q}) - H(\mathbf{t}).$$

By analogy with the definition of $H(\mathbf{p} \mid \mathbf{q})$ as $H(\mathbf{t}) - H(\mathbf{q})$ we can define $H(\mathbf{q} \mid \mathbf{p})$ to be $H(\mathbf{t}) - H(\mathbf{p})$. Then the quantity to be maximized is

$$H(\mathbf{q}) - H(\mathbf{q} \mid \mathbf{p}).$$

We shall describe an alternative way of calculating $H(\mathbf{q} \mid \mathbf{p})$. According to Lemma 5.4, for each input symbol i the numbers $\varGamma_{ij}$ $(j \in J)$ define a probability distribution on the set of output symbols. The entropy of this distribution is a measure of the uncertainty about the output symbol, given that the input symbol is i, and we shall denote it by

$$H(\mathbf{q} \mid i) = \sum_{j \in J} \Gamma_{ij} \log(1/\Gamma_{ij}).$$

The next theorem proves that $H(\mathbf{q} \mid \mathbf{p})$ is the average of these uncertainties, taken over the the set of input symbols.

Theorem 5.13

Let $H(\mathbf{q} \mid \mathbf{p}) = H(\mathbf{t}) - H(\mathbf{p})$. Then

$$H(\mathbf{q} \mid \mathbf{p}) \ = \ \sum_{i \in I} p_i H(\mathbf{q} \mid i).$$

Proof

Let S stand for the sum on the right-hand side. Then we have

$$\begin{aligned} S + H(\mathbf{p}) \ &= \sum_i p_i H(\mathbf{q} \mid i) + \sum_i p_i \log(1/p_i) \\ &= \sum_i p_i \big(H(\mathbf{q} \mid i) + \log(1/p_i) \big). \end{aligned}$$

Using the definition of $H(\mathbf{q} \mid i)$ and the fact that $\sum_j \Gamma_{ij} = 1$, it follows that

$$\begin{aligned} H(\mathbf{q} \mid i) + \log(1/p_i) \ &= \sum_j \Gamma_{ij} \log(1/\Gamma_{ij}) + \sum_j \Gamma_{ij} \log(1/p_i) \\ &= \sum_j \Gamma_{ij} \log(1/p_i \Gamma_{ij}). \end{aligned}$$

Since $p_i \Gamma_{ij} = t_{ij}$, this simplifies to

$$\frac{1}{p_i} \sum_j t_{ij} \log(1/t_{ij}).$$

Hence

$$S + H(\mathbf{p}) = \sum_{i,j} t_{ij} \log(1/t_{ij}) = H(\mathbf{t}),$$

and so $S = H(\mathbf{t}) - H(\mathbf{p}) = H(\mathbf{q} \mid \mathbf{p})$, as claimed. □

Using this result, we can calculate the capacity γ using the alternative formula

$$\gamma = \max_{\mathcal{P}} \Big(H(\mathbf{q}) - H(\mathbf{q} \mid \mathbf{p}) \Big) \quad \text{where } \mathbf{q} = \mathbf{p}\Gamma.$$

Example 5.14

Find the capacity of a channel represented by a matrix of the form

$$\begin{pmatrix} s & t & t & s \\ t & s & s & t \end{pmatrix},$$

where $2(s+t) = 1$.

Solution The capacity is the maximum of $H(\mathbf{q}) - H(\mathbf{q} \mid \mathbf{p})$, where

$$H(\mathbf{q} \mid \mathbf{p}) = p_1 H(\mathbf{q} \mid 1) + p_2 H(\mathbf{q} \mid 2).$$

The entropies on the right-hand side can be calculated directly from the matrix Γ:

$$H(\mathbf{q} \mid 1) = H(\mathbf{q} \mid 2) = 2(s \log(1/s) + t \log(1/t)).$$

Since $p_1 + p_2 = 1$, $H(\mathbf{q} \mid \mathbf{p})$ has the same value. This value is independent of $\mathbf{q}$, so the capacity is obtained when $H(\mathbf{q})$ is a maximum, that is, when $\mathbf{q} = (\frac{1}{4}, \frac{1}{4}, \frac{1}{4}, \frac{1}{4})$, and $H(\mathbf{q}) = \log 4$. Hence the capacity is

$$\log 4 - 2(s \log(1/s) + t \log(1/t)) = 2(1 + s \log s + t \log t).$$

EXERCISES

5.15. A channel Γ accepts input symbols $1, 2, \ldots, 2N$ and produces the output 0 when the input is an even number, and 1 when the input is an odd number. Write down the matrix that represents this channel. Show that for any input distribution $\mathbf{p}$

$$H(\mathbf{p}) - H(\Gamma; \mathbf{p}) = H(\mathbf{q}),$$

where $\mathbf{q} = \mathbf{p}\Gamma$. Hence show that the capacity of this channel is 1 and explain this result.

5.16. Consider a binary asymmetric channel with matrix

$$\Gamma = \begin{pmatrix} 1-a & a \\ b & 1-b \end{pmatrix}.$$

Let the input and output sources be $\mathbf{p}$ and $\mathbf{q} = [q, 1-q]$. Show that $f_\Gamma(\mathbf{p})$ can be written as a function of q in the form

$$F(q) = h(q) - (qx_1 + (1-q)x_2),$$

where x_1 and x_2 satisfy the equation

$$\Gamma \begin{pmatrix} x_1 \\ x_2 \end{pmatrix} = \begin{pmatrix} h(a) \\ h(b) \end{pmatrix}.$$

Noting that x_1 and x_2 do not depend on the source, deduce that the capacity of the channel is $\log_2(2^{x_1} + 2^{x_2})$.

5.17. A simplified keypad has four keys arranged in two rows of two (compare Figure 5.3). If the intention is to press key x, there is probability α of pressing the other key in the same row and probability α of pressing the other key in the same column (and consequently probability $1-2\alpha$ of pressing x). Write down the channel matrix for this situation and find its capacity.

5.18. In Section 5.3 we gave the formulae for deriving p_i, q_j, and Γ_{ij} from the joint distribution $\mathbf{t}$. Verify that, for each $j \in J$, the formula

$$\Delta_{ij} = t_{ij}/q_j = \Pr(\text{input is } i \mid \text{output is } j)$$

defines a probability distribution on I. The entropy of this distribution is denoted by $H(\mathbf{p} \mid j)$. Using the same method as in the proof of Theorem 5.13, show that

$$H(\mathbf{p} \mid \mathbf{q}) = \sum_j q_j H(\mathbf{p} \mid j).$$

[This is often used as the definition of $H(\mathbf{p} \mid \mathbf{q})$. The numbers Δ_{ij} are known as *inverse probabilities*.]

Further reading for Chapter 5

The concepts of a channel and its capacity were formulated by Shannon in his fundamental paper, referenced at the end of Chapter 3 [**3.4**]. This paper was later republished as a book, with a useful introduction by Weaver [**5.3**].

There are several books that develop the theory at a more abstract level, including those by Ash [**5.1**] and McEliece [**5.2**].

5.1 R. Ash. *Information Theory.* Wiley, New York (1965).

5.2 R.J. McEliece. *The Theory of Information and Coding.* Addison-Wesley, Reading, Mass. (1977).

5.3 C.E. Shannon and W. Weaver. *The Mathematical Theory of Communication.* University of Illinois Press, Urbana (1963).

6
The problem of reliable communication

6.1 Communication using a noisy channel

In this chapter we consider the transmission of coded messages through a noisy channel. This is the setting for some very significant results. The general situation is quite complex, and for clarity of exposition we shall consider messages coded in the binary alphabet $\mathbb{B} = \{0, 1\}$, with the symbols being transmitted through a binary symmetric channel. However, it is worth noting that the main results (to be discussed in the next chapter) also hold more generally.

We consider a source emitting a stream of symbols belonging to an alphabet X. The symbols will be encoded as binary words, all having the same length n, where n is going to be chosen so that some desirable conditions are satisfied. For example, suppose the source is a controller guiding a robot through a network of streets, running North-South or East-West. Some streets are blocked by parks and lakes, and some are one-way. In order to move the robot from one location to another, the controller must send a sequence of the symbols N, S, E, W. Each time the robot arrives at a junction it will use the relevant symbol to decide which direction to take (see Figure 6.1).

Here the alphabet is $X = \{\text{N, S, E, W}\}$. Taking $n = 2$, a suitable code $c : X \to \mathbb{B}^2$ would be

$$\text{N} \mapsto 00, \quad \text{S} \mapsto 01, \quad \text{E} \mapsto 10, \quad \text{W} \mapsto 11.$$

Using this code the sequence of instructions shown in Figure 6.1 is encoded as follows:

N. L. Biggs, *An Introduction to Information Communication and Cryptography*,
DOI: 10.1007/978-1-84800-273-9_6, © Springer-Verlag London Limited 2008

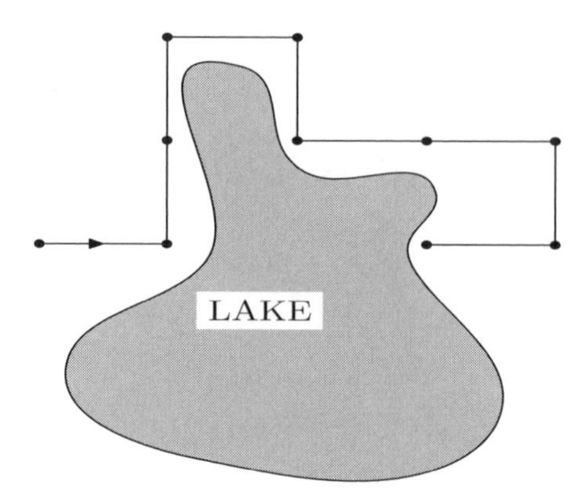

Figure 6.1 The route E,N,N,E,S,E,E,S,W

$$(\mathrm{E}, \mathrm{N}, \mathrm{N}, \mathrm{E}, \mathrm{S}, \mathrm{E}, \mathrm{E}, \mathrm{S}, \mathrm{W}) \mapsto 100000100110100111 \quad .$$

The value $n = 2$ is clearly the smallest possible value for a binary code with four messages. However, this code has the disadvantage that if any error whatsoever is made when the bits are transmitted, the robot will not be able to detect it. For example, suppose the intended symbol is S, so that 01 is sent, and the first bit is altered in transmission, so that 11 is received. The robot will decode this as W, and has no means of knowing that a mistake has been made.

The example illustrates the following scenario. Messages originate from some real source and are expressed in an alphabet X. We shall refer to the output from this source as the *original stream*. A *Sender* must transmit these messages to a *Receiver*, using a binary symmetric channel. In order to do this the Sender encodes the original stream using a binary code $c : X \to \mathbb{B}^n$. We shall refer to the stream emitted by the Sender as the *encoded stream*. The encoded stream is a string of codewords, each of which is a binary word of length n, and so it is in fact a string of bits. In summary, the Sender has effected a transformation

$$\text{T1:} \quad \text{original stream} \longrightarrow \text{encoded stream.}$$

Note that the encoded stream is based on a code which is prefix-free, since all codewords have the same length n. Thus, at this stage, the decoding problem is simple. However, when the bits are transmitted through a BSC, errors are made. The output of the channel, which we shall refer to as the *received stream*, is not the same as the encoded stream. In other words the process of transmission has effected a transformation

$$\text{T2:} \quad \text{encoded stream} \longrightarrow \text{received stream.}$$

We may assume that the Receiver has a 'codebook', so that the set of codewords C and the word-length n are known. But when the Receiver splits the received stream into words of length n, some of the words are not in C, due to bit-errors. Indeed, the received stream may contain any word $z \in \mathbb{B}^n$. In order to understand the message, the Receiver must first decide which codeword $c \in C$ was sent when z is received.

Definition 6.1 (Decision rule)

Let $C \subseteq \mathbb{B}^n$ be a code. A *decision rule* for C is a function $\sigma : \mathbb{B}^n \to C$ which assigns to each $z \in \mathbb{B}^n$ a codeword $c \in C$.

Using a decision rule σ, the Receiver produces a *final stream*, and has effected a transformation

$$\text{T3:} \quad \text{received stream} \longrightarrow \text{final stream.}$$

The entire system is illustrated in Figure 6.2.

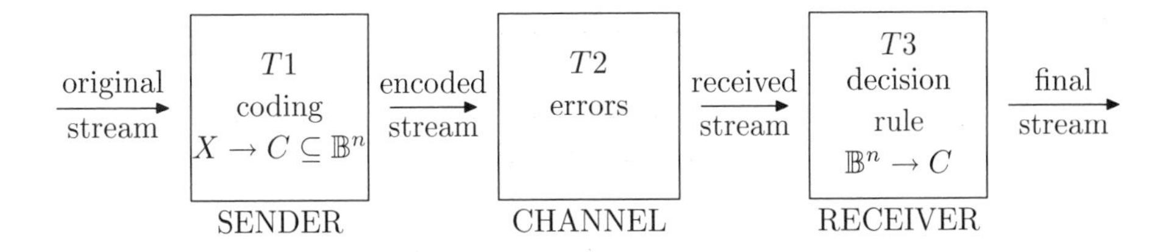

Figure 6.2 The three stages of the communication system

Definition 6.2 (Mistake)

We say that a *mistake* occurs if a codeword in the final stream is not the same as the codeword in the corresponding position in the encoded stream.

When a mistake occurs the Receiver will mis-interpret the message that was sent by the Sender. Mistakes are caused by the errors introduced at stage T2, transmission through a noisy channel. Our aim is to show that, by using an appropriate choice of code C at stage T1, and a suitable decision rule for C at stage T3, it is possible to make mistakes less likely.

For example, consider the scenario described above, where the Sender is a controller emitting the symbols N,S,E,W, using the simple code N $\mapsto$ 00, S $\mapsto$ 01, E $\mapsto$ 10, W $\mapsto$ 11. The Receiver splits the received stream into words of length 2, and in this case all four possible words of length 2 are codewords ($C = \mathbb{B}^2$). Provided that the codewords are transmitted correctly, mistakes will not occur when the Receiver uses the obvious decision rule

$$\sigma(z) = z \quad \text{for all } z \in \mathbb{B}^2.$$

On the other hand, as we have already noted, if any bit is transmitted wrongly then a mistake is certain to occur. Fortunately it is possible to choose better codes.

Example 6.3

In the situation described above, suppose the Sender uses the code

$$\mathrm{N} \mapsto 000, \quad \mathrm{S} \mapsto 110, \quad \mathrm{E} \mapsto 101, \quad \mathrm{W} \mapsto 011,$$

and the Receiver uses the decision rule

$$\sigma(000) = 000, \quad \sigma(100) = 011, \quad \sigma(010) = 000, \quad \sigma(001) = 101,$$
$$\sigma(110) = 110, \quad \sigma(101) = 101, \quad \sigma(011) = 011, \quad \sigma(111) = 110.$$

If a codeword is transmitted with no bit-errors, will a mistake occur? If the word 001 is received and it is assumed that one bit-error has been made, is it possible that a mistake will occur?

Solution The proposed decision rule is such that, when z is codeword, $\sigma(z) = z$. Thus if a codeword is transmitted correctly, a mistake will not occur.

What happens if one bit in a codeword is transmitted wrongly? Note that each codeword contains an even number of 1's (either 0 or 2). Thus if there is one erroneous bit the received word will have an odd number of 1's, and the error can be *detected*. However, if 001 is received (for example), then the Receiver must decide whether the transmitted codeword was 000, with an error in the last bit, or 101, with an error in the first bit, or 011, with an error in the second bit. The proposed decision rule assumes the second possibility, and so if 001 is received then there is a significant chance that a mistake will occur. In fact, there are other reasons why this rule is rather a poor one (see Section 6.3).

Example 6.4

In the same situation, suppose the Sender uses the code

$$\mathrm{N} \mapsto 000000, \quad \mathrm{S} \mapsto 000111, \quad \mathrm{E} \mapsto 111000, \quad \mathrm{W} \mapsto 111111.$$

The Receiver uses the decision rule that for any z, $\sigma(z) = c$ is the codeword that is 'most like' z – that is, the codeword c for which z and c have most bits in common. In what circumstances will a mistake occur?

Solution The codewords have been chosen so that if any one bit is changed, the resulting word is still 'more like' the original codeword than any other codeword. The proposed decision rule makes use of this property. For example, if 000000 is altered in transmission to 100000 then the Receiver will decide that 000000 was intended, because any other codeword would have to be affected by more than one bit-error to produce 100000. So an error in one bit is not only detected, but also *corrected.* For a mistake to occur, at least two bit-errors must be made.

The system outlined above is a complicated one, involving several stages. In due course we shall be able to explain how the objective of making mistakes less likely can be achieved, but first we must study the various stages of the system in detail.

- Section 6.2 quantifies the errors that occur when the bits in a codeword of length n are transmitted through a BSC with bit-error probability e.
- Section 6.3 discusses the possible forms of a decision rule σ that assigns a codeword to each (possibly erroneous) received word.
- Sections 6.4 and 6.5 deal with the correction of bit-errors, and how it depends on certain numerical parameters of the code that is used.

EXERCISES

6.1. In Example 6.3, suppose that 111 is received. If it is assumed that only one bit is in error, what are the possibilities for the intended instruction?

6.2. In Example 6.4 we proposed a decision rule σ based on the idea that $\sigma(z)$ is the codeword that is 'more like' z than any other codeword. Using this rule, find the values of $\sigma(z)$ for the following words z:

$$101000, \qquad 101111, \qquad 100111.$$

6.3. In the same scenario as in Exercise 6.2, suppose that the word 100100 is received. Is it possible to make a reasonable decision about which codeword was sent?

6.2 The extended BSC

Definition 6.5 (Product channel)

Let Γ and Γ' be channels with input alphabets I, I' and output alphabets J, J' respectively. The *product channel* $\Gamma'' = \Gamma \times \Gamma'$ has input alphabet $I \times I'$ and output alphabet $J \times J'$, and its matrix is given by the rule

$$(\Gamma'')_{ii'\, jj'} = \Gamma_{ij}\Gamma'_{i'j'}.$$

In other words Γ'' is a channel with inputs ii' and outputs jj', and

$$\Pr(\text{output is } jj' \mid \text{input is } ii') = \Pr(j \mid i)\Pr(j' \mid i').$$

In the case $\Gamma = \Gamma'$, we denote $\Gamma \times \Gamma'$ by Γ^2. (Note that this does *not* correspond to the usual product of matrices.)

Example 6.6

Suppose Γ is the BSC with bit-error probability e. What is the channel matrix for Γ^2?

Solution The inputs and outputs for Γ^2 are the four pairs $00, 01, 10, 11$. If, for example, the input is 00 and the output is 10, this means that one bit has been transmitted wrongly (probability e), and one bit has been transmitted correctly (probability $1-e$). So the entry in the corresponding position in Γ^2 is $e(1-e)$.

Using similar arguments, and labelling the rows and columns in the order $00, 01, 10, 11$, the complete channel matrix is

$$\begin{pmatrix} (1-e)^2 & e(1-e) & e(1-e) & e^2 \\ e(1-e) & (1-e)^2 & e^2 & e(1-e) \\ e(1-e) & e^2 & (1-e)^2 & e(1-e) \\ e^2 & e(1-e) & e(1-e) & (1-e)^2 \end{pmatrix}.$$

Definition 6.7 (Extended channel)

Suppose we are given a channel Γ with input set I and output set J, and a positive integer n. Then we define the *extended channel* Γ^n to be the n-fold product of copies of Γ. The inputs to Γ^n are words of length n in I, and the outputs are words of length n in J. If Γ is a binary symmetric channel, then we say that Γ^n is an *extended BSC*.

The inputs and outputs for an extended BSC are the members of the set $\mathbb{B}^n$ of all binary words of length n. We shall obtain a simple formula for the entries of the channel matrix Γ^n, by generalizing the argument used above in the case $n = 2$. The following definition is crucial.

Definition 6.8 (Hamming distance)

Suppose we are given two words $x, y \in \mathbb{B}^n$,

$$x = x_1 x_2 \cdots x_n \qquad y = y_1 y_2 \cdots y_n.$$

The *Hamming distance* $d(x, y)$ is the number of places where x and y differ; in other words, it is the number of i ($1 \leq i \leq n$) such that $x_i \neq y_i$.

For example, consider the following words in $\mathbb{B}^7$:

$$x = 1010100, \qquad y = 0110100, \qquad z = 1011110.$$

The words x and y differ in the first and second bits only, so $d(x, y) = 2$. Similarly $d(x, z) = 3$ and $d(y, z) = 4$.

Theorem 6.9

Let $x, y \in \mathbb{B}^n$. The entry $(\Gamma^n)_{xy}$ in the channel matrix for the extended BSC with bit-error probability e is given by

$$(\Gamma^n)_{xy} = e^d (1 - e)^{n-d}, \text{ where } d = d(x, y).$$

Proof

Suppose that $x \in \mathbb{B}^n$ is sent and $y \in \mathbb{B}^n$ is received. If $d(x, y) = d$, then d bits are in error, and the probability of this event is e^d. The remaining $n - d$ bits are correct, and the probability of this event is $(1 - e)^{n-d}$. Hence

$$(\Gamma^n)_{xy} = \Pr(y \mid x) = e^d (1 - e)^{n-d}.$$

□

EXERCISES

6.4. Let Γ^2 be the channel matrix for the extended BSC with $e = 0.01$. Calculate the numbers in the row of Γ^2 corresponding to the input 00.

6.5. Write down the channel matrix for Γ^2 when Γ is the binary *asymmetric* channel with parameters a, b (Exercise 5.1).

6.6. Let d denote the Hamming distance. Prove that, for all $x, y, z \in \mathbb{B}^n$,

$$d(x,x) = 0, \qquad d(x,y) = d(y,x), \qquad d(x,z) \leq d(x,y) + d(y,z).$$

[The last property is known as the *triangle inequality*.]

6.3 Decision rules

In Section 6.1 we introduced the idea of a *decision rule* σ for a code $C \subseteq \mathbb{B}^n$. The idea is that when a word $z \in \mathbb{B}^n$ is received, the Receiver must try to make a reasonable choice as to which codeword $\sigma(z) \in C$ was sent. As an example we considered the code

$$C = \{000, 110, 101, 011\},$$

and the arbitrary decision rule σ given by

$$\sigma(000) = 000, \quad \sigma(100) = 011, \quad \sigma(010) = 000, \quad \sigma(001) = 101,$$

$$\sigma(110) = 110, \quad \sigma(101) = 101, \quad \sigma(011) = 011, \quad \sigma(111) = 110.$$

Since the word 011 is actually a codeword, it is reasonable to define $\sigma(011) = 011$. On the other hand, 100 is not codeword, and putting $\sigma(100) = 011$ assumes that an error has been made in each of the three bits, which is clearly unreasonable.

A good decision rule will depend on the communication system, in particular, the code C and the channel matrix Γ. The Receiver must use this information to formulate a decision rule that provides the best chance of making the right choice. At first sight, the obvious candidate is the following.

Definition 6.10 (Ideal observer rule)

The *ideal observer rule* says that, given z, the Receiver should choose $\sigma(z)$ to be a codeword c for which the probability that c was sent, given that z is received, is greatest.

Unfortunately, the conditional probabilities occurring in this definition are of the form $\Pr(c \mid z)$, and these are not necessarily available to the Receiver.

In order to calculate $\Pr(c \mid z)$ it is necessary to equate two expressions for the probability t_{cz} of an input-output pair (c, z):

$$t_{cz} = q_z \Pr(c \mid z) \qquad t_{cz} = p_c \Pr(z \mid c).$$

By definition $\Pr(z \mid c)$ is just the term Γ_{cz} in the channel matrix. However $\Pr(c \mid z)$ must be obtained from the equation $\Pr(c \mid z) = p_c \Gamma_{cz}/q_z$, and this involves the input distribution $\mathbf{p}$ as well as Γ. In general the characteristics of the input will not be known to the Receiver – for example, when the original stream is generated by some remote-sensing device.

However, it is reasonable to assume that the Receiver knows the characteristics of the channel, in particular the probabilities $\Gamma_{cz} = \Pr(z \mid c)$. Thus a more practical rule is the following.

Definition 6.11 (Maximum likelihood rule)

The *maximum likelihood rule* says that, given z, the Receiver should choose $\sigma(z)$ to be a codeword c that satisfies

$$\Pr(z \mid c) \geq \Pr(z \mid c') \quad \text{for all } c' \in C.$$

The maximum likelihood rule says that $\sigma(z)$ should be a codeword c for which the probability that z is received, given that c is sent, is greatest. Writing the condition as $\Gamma_{cz} \geq \Gamma_{c'z}$ we see that the rule can be expressed as follows. Given z, pick a largest term Γ_{cz} in column z of the channel matrix and put $\sigma(z) = c$. Note that there may be more than one c satisfying this condition, in which case the receiver must make an arbitrary choice.

When the channel is the extended BSC we have a formula (Theorem 6.9) for the terms of the channel matrix, and this leads to a very simple interpretation of the maximum likelihood rule.

Definition 6.12 (Minimum distance rule)

The *minimum distance rule* (or *MD rule*) says that given $z \in \mathbb{B}^n$, the Receiver should choose $\sigma(z)$ to be a codeword c such that $d(z, c)$ is a minimum.

Theorem 6.13

For an extended BSC with $e < \frac{1}{2}$, the maximum likelihood rule is equivalent to the MD rule.

Proof

Suppose that c is codeword such that $d(z, c) = i$. Then, according to Theorem 6.9, $\Pr(z \mid c) = e^i(1-e)^{n-i}$. If c' is codeword such that $d(z, c') = i'$, then

$$\frac{\Pr(z \mid c)}{\Pr(z \mid c')} = \frac{e^i(1-e)^{n-i}}{e^{i'}(1-e)^{n-i'}} = \left(\frac{1-e}{e}\right)^{i'-i}.$$

If $e < \frac{1}{2}$ then $(1-e)/e > 1$. Thus when $i < i'$ the right-hand side is greater than 1, and $\Pr(z \mid c) > \Pr(z \mid c')$. In other words, choosing c so that $d(z, c)$ is minimal is the same as maximizing $\Pr(z \mid c)$. □

The theorem says that the Receiver can implement the maximum likelihood rule by choosing as $\sigma(z)$ a codeword that is nearest to z, in the sense of Hamming distance. For some words z there will be more than one such codeword, and a choice will be needed. There are several ways in which the choice can be made. The simplest is for the Receiver to select one of the eligible codewords once and for all, and declare it to be $\sigma(z)$. Alternatively the Receiver may choose one of the eligible codewords uniformly at random each time z occurs. In other words, if there are k eligible codewords, each one is chosen as $\sigma(z)$ with probability $1/k$. The second alternative seems more reasonable than the first, but from the theoretical point of view it creates some difficulties, because σ is no longer strictly a function (see Exercise 6.9).

We shall continue to speak of 'the MD rule', remembering that in practice there may be more than one MD rule for any given situation.

Example 6.14

Suppose the code $C \subseteq \mathbb{B}^8$ has seven codewords

$$c_1 = 0000\,0000 \quad c_2 = 0011\,1000 \quad c_3 = 1100\,0001 \quad c_4 = 0000\,1110$$

$$c_5 = 1011\,1011 \quad c_6 = 0011\,0110 \quad c_7 = 1100\,1011.$$

If σ is the MD rule, what value should the Receiver assign to

(i) $\sigma(1010\,1011)$ (ii) $\sigma(1100\,1001)$?

Solution (i) The distances $d(z, c_i)$ for $z = 1010\,1011$ are

i :	1	2	3	4	5	6	7
$d(z, c_i)$:	5	4	4	4	1	5	2

Hence the codeword nearest to z is $c_5 = 1011\,1011$. According to the theorem, this is also the codeword c for which the probability of receiving z, given that c was sent, is greatest.

(ii) The distances $d(z', c_i)$ for $z' = 1100\ 1001$ are

$$\begin{array}{lccccccc} i: & 1 & 2 & 3 & 4 & 5 & 6 & 7 \\ d(z', c_i): & 4 & 5 & 1 & 5 & 4 & 8 & 1 \end{array} .$$

In this case there are two codewords c_3 and c_7 that could be chosen as $\sigma(z')$. The Receiver may either fix on one of them, or whenever z' occurs, choose one of them at random (say by tossing a fair coin).

EXERCISES

6.7. Let $C \subseteq \mathbb{B}^8$ be the code discussed in Example 6.14, with codewords

$$c_1 = 0000\ 0000 \quad c_2 = 0011\ 1000 \quad c_3 = 1100\ 0001 \quad c_4 = 0000\ 1110$$

$$c_5 = 1011\ 1011 \quad c_6 = 0011\ 0110 \quad c_7 = 1100\ 1011.$$

Using the MD rule σ, find $\sigma(z)$ when z is

$$1000\ 1011, \qquad 1011\ 1010, \qquad 1100\ 0101.$$

6.8. Consider the code C consisting of the 10 words in $\mathbb{B}^5$ that have exactly two bits equal to 1. If one bit-error is made in transmitting the codeword 11000, what are the possible received words z? For each such z, make a list of the codewords that are nearest to z.

6.9. Under the same conditions as in the previous exercise, suppose the Receiver uses the MD rule, with the proviso that if there are k codewords that are nearest to a given word z, one of them is chosen as $\sigma(z)$ with probability $1/k$. Suppose the codeword 11000 is sent, and probability of a single bit-error in transmission is e. Show that there is probability $7e/12$ that the Receiver will decide (wrongly) that the codeword sent was 10001.

6.10. Let Γ be a binary asymmetric channel with property that 0 is always transmitted correctly, but 1 is transmitted as 0 with probability f $(0 < f < 1)$. Write down the entries $(\Gamma^3)_{cz}$ of the extended channel matrix that correspond to codewords in the code $C = \{000, 111\}$. What is the maximum likelihood decision rule for C, using this channel?

6.11. Show that the maximum likelihood rule is not the same as the MD rule for the channnel considered in the previous exercise.

6.4 Error correction

The purpose of a decision rule is to ensure that, as far as possible, bit-errors are corrected. We shall focus on the MD rule for the extended BSC, where the effectiveness of the rule depends upon certain parameters of the code. The reader will find it helpful to envisage the following discussion in geometrical terms.

Definition 6.15 (Minimum distance)

Let $d(c, c')$ denote the Hamming distance between codewords c and c' in a code $C \subseteq \mathbb{B}^n$ (Definition 6.8). The *minimum distance* of C is

$$\delta = \min_{c \neq c'} d(c, c').$$

For example, suppose $C \subseteq \mathbb{B}^6$ has four codewords

$$000000,\ 111000,\ 001110,\ 110011.$$

The distances between the codewords are as follows:

$$d(000000, 111000) = 3, \quad d(000000, 001110) = 3, \quad d(000000, 110011) = 4,$$

$$d(111000, 001110) = 4, \quad d(111000, 110011) = 3, \quad d(001110, 110011) = 5,$$

and so the minimum distance is $\delta = 3$.

Definition 6.16 (Neighbourhood)

For any word $x \in \mathbb{B}^n$ and any $r \geq 0$, the *neighbourhood* of x with radius r is the set

$$N_r(x) = \{y \in \mathbb{B}^n \mid d(x, y) \leq r\}.$$

Equivalently, the neighbourhood $N_r(x)$ contains all the words that can be obtained from x by making not more than r bit-errors. For example, take $x = 11010 \in \mathbb{B}^5$. Then the neighbourhood $N_1(x)$ contains x itself and the five words obtained by making one error in x, that is

$$N_1(x) = \{11010, 01010, 10010, 11110, 11000, 11011\}.$$

Lemma 6.17

If C is a code with $\delta \geq 2r + 1$, then for any codewords $c, c' \in C$, the neighbourhoods $N_r(c)$ and $N_r(c')$ are disjoint (see Figure 6.3).

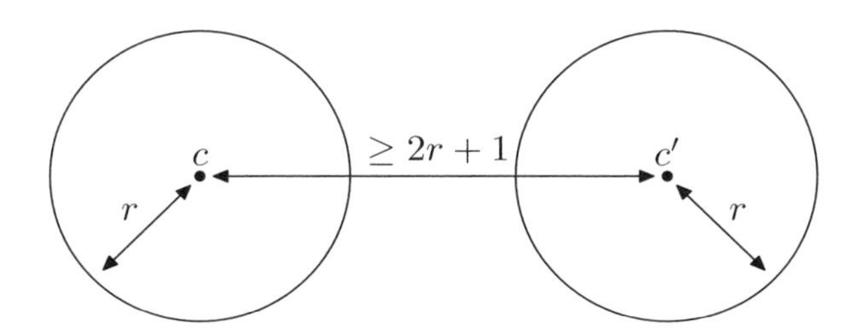

Figure 6.3 If $\delta \geq 2r+1$ then $N_r(c)$ and $N_r(c')$ are disjoint

Proof

Suppose $N_r(c)$ and $N_r(c')$ are not disjoint, so they have a common member z, say. Then by the triangle inequality (Exercise 6.6),

$$d(c, c') \leq d(c, z) + d(z, c') \leq r + r = 2r.$$

But this contradicts the assumption that $\delta \geq 2r+1$, since $d(c, c')$ cannot be less than δ. □

Theorem 6.18

Suppose the Sender uses a code C having minimum distance $\delta \geq 2r+1$, and the Receiver uses the MD rule. Then, provided that not more than r bit-errors are made in transmitting any codeword, no mistakes will occur: every received word will be restored to the correct codeword.

Proof

Let c be a codeword, and z the corresponding received word. If not more than r bit-errors are made in transmission, then z is in $N_r(c)$. According to the lemma, z is not in $N_r(c')$ for any $c' \neq c$. In other words, $d(c, z) < d(c', z)$ for all $c' \neq c$. This means that the MD rule will correctly assign c to z. □

Definition 6.19 (Error-correcting code)

A code $C \subseteq \mathbb{B}^n$ is *an r-error-correcting code* if $\delta \geq 2r+1$.

The definition is based on the implicit assumption that the MD rule is used. We sometimes say (even more vaguely) that a code with $\delta \geq 2r+1$ 'corrects r errors'. For example, if δ is at least 3, then C corrects 1 error.

EXERCISES

6.12. Construct a 1-error correcting code $C \subseteq \mathbb{B}^6$ with $|C| = 5$.

6.13. For the code C constructed in the previous exercise, how many words cannot be corrected on the assumption that at most one bit-error has been made?

6.14. Is the code with $|C| = 5$ constructed in Exercise 6.12 the largest possible 1-error-correcting binary code with words of length 6?

6.5 The packing bound

When we try to construct error-correcting codes by ensuring that δ has a given value, we run into a problem. In geometrical terms, it is intuitively clear that choosing a set of codewords C in $\mathbb{B}^n$, such that the codewords are 'far apart' (in the sense of Hamming distance), means that the number of codewords $|C|$ is constrained. The following theorem quantifies this remark.

Theorem 6.20 (The packing bound)

If $C \subseteq \mathbb{B}^n$ is a code with $\delta \geq 2r + 1$, then

$$|C|\left(1 + n + \binom{n}{2} + \cdots + \binom{n}{r}\right) \leq 2^n.$$

Proof

Given a codeword c, the words z such that $d(c, z) = i$ are the words formed by altering any i of the n bits in c. Hence the number of such words is the binomial number $\binom{n}{i}$. The number of z such that $d(c, z) \leq r$ is thus

$$1 + n + \binom{n}{2} + \cdots + \binom{n}{r},$$

and this is the size of the neighbourhood $N_r(c)$ for each $c \in C$.

Since $\delta \geq 2r + 1$, it follows from Lemma 6.17 that all the neighbourhoods $N_r(c)$ $(c \in C)$ are disjoint (Figure 6.4). There are $|C|$ such neighbourhoods, and 2^n words altogether, so the result follows. □

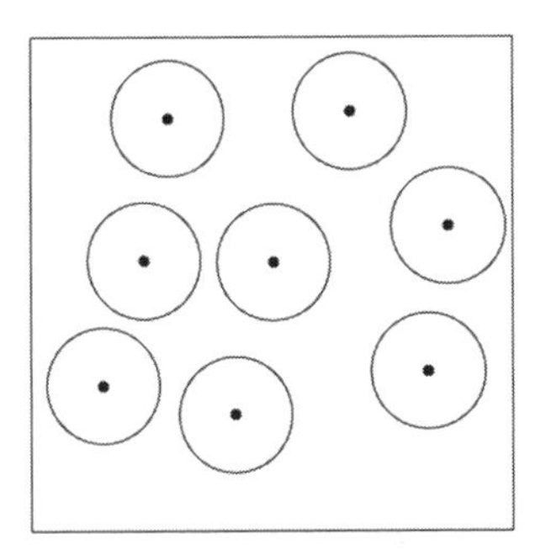

Figure 6.4 If $\delta \geq 2r+1$ all the neighbourhoods $N_r(c)$ are disjoint

Example 6.21

Suppose we want to issue ID numbers, in the form of strings of n bits, to 100 students. The students are likely to make errors when using their ID's, so it has been decided that we must use a 2-error-correcting code. What is the least value of n for which this may be possible?

Solution For $r = 2$ errors and $|C| = 100$ the packing bound is

$$100\left(1+n+\frac{n(n-1)}{2}\right) \leq 2^n, \qquad \text{that is}$$

$$50(n^2+n+2) \leq 2^n.$$

By trial and error, the least value of n for which this inequality holds is $n = 14$.

Of course, the problem of constructing a suitable set of 100 codewords, each of length 14, remains.

Let us denote by $A(n, \delta)$ the largest value of $|C|$ for which there is a code $C \subseteq \mathbb{B}^n$ with minimum distance δ. The packing bound gives an upper limit for $A(n, \delta)$ but there is no guarantee that a code of that size actually exists. Indeed the problem of evaluating $A(n, \delta)$ exactly is, in general, very difficult.

Example 6.22

What is the value of $A(10, 7)$, the largest possible size of a code $C \subseteq \mathbb{B}^{10}$ with minimum distance $\delta = 7$?

Solution Here we require $\delta = 2r+1$ with $r = 3$, so the packing bound is

$$|C|\left(1+10+\binom{10}{2}+\binom{10}{3}\right) \leq 2^{10},$$

that is, $|C| \leq 1024/176$. Since $|C|$ is an integer, it follows that $A(10,7) \leq 5$.

In fact $A(10,7) = 2$. To prove this, we can suppose without loss of generality that one codeword is 0000000000. Then any other codeword must have at least 7 ones. Now two words of length 10 with at least 7 ones must have at least 4 ones in the same positions. Hence if these words are distinct, their distance can be at most 6, contrary to the condition $\delta = 7$. It follows that there can be only one other codeword besides 0000000000.

There is another, very useful, way of interpreting the constraint imposed by the packing bound. When $C \subseteq \mathbb{B}^n$ and $|C| = m$, a codeword $c \in C$ conveys (at most) $\log_2 m$ bits of information, but requires n bits of data. Hence the following definition.

Definition 6.23 (Information rate)

The *information rate* of a code $C \subseteq \mathbb{B}^n$ is

$$\rho = \frac{\log_2 |C|}{n}.$$

For example, suppose $C \subseteq \mathbb{B}^6$ has four codewords

$$000000,\ 111000,\ 001110,\ 110011.$$

Here $n = 6$ and $|C| = 4 = 2^2$, so the information rate is $\rho = 2/6 = 1/3$. This corresponds to the fact that C requires 6 bits of data to encode 2 bits of information. If we suppose that one megabyte of data can be transmitted per second then, when C is used to code the data, only one-third of a megabyte of information is actually transmitted per second.

Constructing a code $C \subseteq \mathbb{B}^n$ with given size $|C|$ and a given value of δ is equivalent to constructing a code with given values of n, ρ, and δ. The packing bound tells us that there is a 'trade-off' between δ and ρ.

EXERCISES

6.15. If we require a code $C \subseteq \mathbb{B}^6$ with information rate at least 0.35, what is the smallest possible value of $|C|$?

6.16. Show that $(\log_2 9)/6 \approx 0.528$ is an upper bound for the information rate of a 1-error-correcting code $C \subseteq \mathbb{B}^6$.

6.17. Suppose we want to issue ID numbers, in the form of strings of n bits, to 1000 people, using a 1-error-correcting code. What is the least value of n for which this may be possible?

6.18. Using a general form of the proof that $A(10, 7) = 2$ (Example 6.22), show that if $\delta > 2n/3$ then $A(n, \delta) = 2$.

6.19. Prove that there is an r-error-correcting code $C \subseteq \mathbb{B}^n$ satisfying

$$|C| \left(1 + n + \binom{n}{2} + \cdots + \binom{n}{2r}\right) \geq 2^n.$$

[This result gives a lower bound for $A(n, 2r + 1)$. It is known as the *Gilbert-Varshamov bound.*]

6.20. An r-error-correcting code $C \subseteq \mathbb{B}^n$ is said to be *perfect* if the neighbourhoods $N_r(c)$ exactly cover $\mathbb{B}^n$ – that is, every word in $\mathbb{B}^n$ is in precisely one $N_r(c)$. Show that if $r = 1$ and such a code exists, then n must be of the form $2^m - 1$. [See also Section 9.1.]

Further reading for Chapter 6

There are many books on error-correcting codes, at various levels. The standard works by MacWilliams and Sloane [**6.2**], and Pless and Huffman [**6.3**] are recommended for a thorough treatment of the subject. A readable account of the early history, and the connections with geometry and algebra, is given by Thompson [**6.4**].

Estimates of the numbers $A(n, \delta)$ are constantly being improved. A up-to-date list is maintained by Litsyn, Rains, and Sloane [**6.1**].

6.1 S. Litsyn, E.M. Rains, N.J.A. Sloane. Table of Nonlinear Binary Codes. http://www.research.att.com/~njas/codes/And/

6.2 F.J. MacWilliams and N.J.A. Sloane. *The Theory of Error-Correcting Codes.* North-Holland, Amsterdam (1977).

6.3 V. Pless and W. Huffman (eds.). *Handbook of Coding Theory* (2 vols). Elsevier, Amsterdam (1998).

6.4 T.M. Thompson. *From Error-Correcting Codes through Sphere Packings to Simple Groups.* Carus Mathematical Monographs 21, Math. Assoc. of America (1983).

7 The noisy coding theorems

7.1 The probability of a mistake

Here again (Figure 7.1) is the model of communication outlined at the start of the previous chapter.

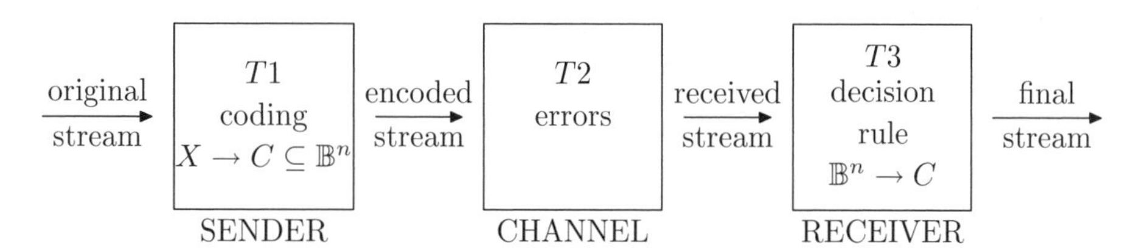

Figure 7.1 A model of a communication system

Suppose that the *original stream* is emitted by a source that produces symbols from an alphabet X, according to the probability distribution $\mathbf{p}$ on X. When the Sender uses the code C, the probability that a codeword $c \in C$ is sent is the same as the probability that the source emits the symbol $x \in X$ for which c is the codeword. Thus we can regard the *encoded stream* as being emitted by a source $(C, \mathbf{p})$. Since $\mathbf{p}$ may be unknown, the appropriate decision rule is the maximum likelihood rule. Given that the channel is the extended BSC, the Receiver can use its equivalent form, the MD rule.

N. L. Biggs, *An Introduction to Information Communication and Cryptography*,
DOI: 10.1007/978-1-84800-273-9_7, © Springer-Verlag London Limited 2008

For each $c \in C$, denote by $F(c)$ the set of words to which the MD rule σ does *not* assign c:

$$F(c) = \{z \in \mathbb{B}^n \mid \sigma(z) \neq c\}.$$

According to Definition 6.2 a *mistake* occurs when a codeword $c \in C$ is altered in transmission and the received word z is in $F(c)$. In this situation, the Receiver will decide that a codeword c' different from c was sent.

For a given codeword c let M_c denote the probability that when c is sent the received word z is in $F(c)$, in other words M_c is the sum of the conditional probabilities $\Pr(z \mid c)$ for $z \in F(c)$:

$$M_c = \sum_{z \in F(c)} \Pr(z \mid c).$$

Definition 7.1 (Probability of a mistake)

The *probability of a mistake* when the encoded stream corresponds to a source $(C, \mathbf{p})$ is

$$M(C, \mathbf{p}) = \sum_{c \in C} p_c M_c,$$

where M_c is given by the formula displayed above.

The aim is to choose the code C so that $M(C, \mathbf{p})$ is as small as we please. The following example illustrates how this might be achieved.

Example 7.2

An investor wishes to communicate the messages *BUY*, *SELL* to her stockbroker. Suppose she uses one of the codes

(i) $BUY \mapsto 0, \quad SELL \mapsto 1;$ (ii) $BUY \mapsto 000, \quad SELL \mapsto 111.$

The encoded messages are transmitted via an extended BSC with bit-error probability e, and the stockbroker uses the MD rule. Which code has the smaller probability of a mistake?

Solution (i) Here $n = 1$ and the code is $C_1 = \{0, 1\}$. The MD rule is $\sigma(0) = 0$, $\sigma(1) = 1$, and the sets $F(c)$ are

$$F(0) = \{1\}, \qquad F(1) = \{0\}.$$

The probabilities M_0 and M_1 are

$$M_0 = \Pr(1 \mid 0) = e, \qquad M_1 = \Pr(0 \mid 1) = e.$$

Hence, for any distribution $\mathbf{p} = [p, 1-p]$ on C_1,

$$M(C_1, \mathbf{p}) = pe + (1-p)e = e.$$

(ii) Here $n = 3$ and the code is $C_3 = \{000, 111\}$. The MD rule is

$$\sigma(000) = \sigma(100) = \sigma(010) = \sigma(001) = 000,$$

$$\sigma(011) = \sigma(101) = \sigma(110) = \sigma(111) = 111.$$

With this rule

$$F(000) = \{011, 101, 110, 111\}, \quad F(111) = \{000, 001, 010, 100\}.$$

If the codeword 000 is sent, the probability M_{000} of a mistake is thus

$$\begin{array}{llll} & \Pr(011 \mid 000) & + \Pr(101 \mid 000) & + \Pr(110 \mid 000) & + \Pr(111 \mid 000) \\ = & e^2(1-e) & + e^2(1-e) & + e^2(1-e) & + e^3. \end{array}$$

A very rough approximation tells us that all four terms are less that e^2, so $M_{000} < 4e^2$. A similar calculation shows that $M_{111} < 4e^2$ also. Hence, for any distribution $\mathbf{p} = [p, 1-p]$ on C_3,

$$M(C_3, \mathbf{p}) < p(4e^2) + (1-p)(4e^2) = 4e^2.$$

Thus, when e is small, $M(C_3, \mathbf{p})$ is much less than $M(C_1, \mathbf{p})$. For example, when $e = 0.001$, we have

$$M(C_1, \mathbf{p}) = 0.001, \qquad M(C_3, \mathbf{p}) < 0.000004.$$

In general, can we give estimates for the probability of a mistake, based on arguments like those used in the example? The following simple remark is a starting point.

Lemma 7.3

If $C \subseteq \mathbb{B}^n$ is an r-error-correcting code, and z is in $F(c)$, then $d(z, c) \geq r + 1$.

Proof

Theorem 6.18 says that if the received word z is such that $d(z, c) \leq r$, then the Receiver will correctly decide that c was sent. Hence the words for which the Receiver decides on the wrong codeword must satisfy $d(z, c) \geq r + 1$. $\square$

When a code $C \subseteq \mathbb{B}^n$ is used we have (Theorem 6.9)

$$\Pr(z \mid c) = (\Gamma^n)_{cz} = e^{d(c,z)}(1-e)^{n-d(c,z)}.$$

Hence

$$M_c = \sum_{z \in F(c)} e^{d(c,z)}(1-e)^{n-d(c,z)}.$$

According to Lemma 7.3 all the summands in M_c are of the form

$$e^i(1-e)^{n-i}, \quad \text{for some} \quad i \geq r+1.$$

Since $0 < e < 1$, we have

$$e^i(1-e)^{n-i} < e^i \leq e^{r+1}.$$

In other words, M_c is a sum of terms of order e^{r+1}. In specific cases we can estimate the size of $F(c)$, and hence obtain an upper bound for M_c. Thus, in Example 7.2, C_3 is a 1-error correcting code, and so each term in M_c is of order e^2. There are four words in $F(c)$, so we can conclude that M_c is bounded above by $4e^2$.

The problem with this approach is that $F(c)$ is not necessarily a small set. This means that the suggested upper bound

$$M_c \leq \sum_{z \in F(c)} e^{r+1} = |F(c)|\, e^{r+1},$$

is such that the 'constant' multiplier of e^{r+1} may be very large. For this reason a more careful analysis of the problem is needed.

EXERCISES

7.1. Let $C = \{00000, 11100, 00111\}$. If σ is an MD decision rule for C, show that $F(c)$ is a set of size 12, at least, for each codeword c.

7.2. Consider the code $C = \{0, 1\}$ as the input to a BSC with bit-error probability $e = 0.2$. Verify that the maximum likelihood decision rule is given by $\sigma(0) = 0$, $\sigma(1) = 1$, and that the probability of a mistake is 0.2, for any distribution $\mathbf{p}$ on C.

7.3. In the previous exercise, suppose the input distribution is $\mathbf{p} = [0.9, 0.1]$, and this is known to the Receiver, who can therefore use the ideal observer rule (Definition 6.10). Write down the rule explicitly, and show that using this rule the probability of a mistake is reduced to 0.1.

7.4. The *repetition code* $R_n \subseteq \mathbb{B}^n$ consists of the two words $000\cdots00$ and $111\cdots11$. If n is an odd number $2\ell+1$, and σ is the MD rule, find the size of the sets $F(000\cdots00)$ and $F(111\cdots11)$.

7.5. Suppose that R_5 is transmitted through an extended BSC with bit-error probability e, and the MD rule is used. Calculate M_{00000}, M_{11111}, and $M(R_5, \mathbf{p})$ exactly as functions of e, for any distribution $\mathbf{p}$ on R_5.

7.6. Show that, provided $e < \frac{1}{4}$, we can make $M(R_{2\ell+1}, \mathbf{p})$ as small as we please by choosing ℓ sufficiently large.

7.7. In Exercise 6.10 we considered a binary asymmetric channel Γ with the property that 0 is always transmitted correctly, but 1 is transmitted as 0 with probability f $(0 < f < 1)$. Suppose the source $(R_n, \mathbf{p})$ is transmitted through the extended channel Γ^n, where $\mathbf{p} = [p, 1-p]$. What is the behaviour of the information rate and the probability of a mistake as $n \to \infty$?

7.2 Coding at a given rate

We continue to discuss the model shown in Figure 7.1. Specifically, stage T1 represents encoding with a binary code C, stage T2 represents transmission through an extended BSC, and stage T3 represents the application of the MD rule.

> *Question:* *Is it possible to choose the code C so that the probability of a mistake is arbitrarily small, while information is transmitted at a given rate ρ?*

We shall see that the answer is YES, provided ρ is not too large. In fact, the key to maintaining a given information rate ρ is to code blocks of symbols, rather than individual symbols.

It is convenient to suppose that the original stream is already in the form of a string of bits. This can be arranged by using a simple 'pre-coding' rule. For example, if the original stream is a sequence of commands N, S, E, W (as in Section 6.1), we could use the obvious pre-coding N $\mapsto$ 00, S $\mapsto$ 01, E $\mapsto$ 10, W $\mapsto$ 11.

Consider the following strategy for stage T1. The Sender divides the original stream of bits into blocks of a certain size, say k, and assigns to each block a codeword belonging to a code C. There are 2^k possible blocks, so a code of size $|C| = 2^k$ will be required. In order to ensure that the information rate is not less than some given value ρ, the Sender must determine the appropriate length n for the codewords. The information rate of the code C is $(\log_2 |C|)/n = k/n$. Hence the Sender must choose the parameters k, n, and the code $C \subseteq \mathbb{B}^n$, such that

$$|C| = 2^k \qquad \text{and} \quad k \geq \rho n.$$

The Sender also wishes to choose the code C so that the probability of a mistake is small, knowing that the encoded stream will be transmitted through an extended BSC with bit-error probability e, and the Receiver will use the MD decision rule. These conditions imply that the trade-off between the parameters ρ and δ will operate.

Example 7.4

Suppose that the desired information rate is $\rho = 0.8$, and the Sender wishes to use a 1-error correcting code ($\delta \geq 3$). What are the smallest possible values of n and k?

Solution If $C \subseteq \mathbb{B}^n$ is a 1-error correcting code with $|C| = 2^k$, it follows from the packing bound (Theorem 6.20) that $2^k(1+n) \leq 2^n$. Also, in order to achieve the rate $\rho = 0.8$, the Sender must ensure that k is not less than $0.8n$. Thus we we require the least value of n for which there is an integer k satisfying

$$n+1 \ \leq \ 2^{n-k} \qquad \text{and} \qquad 5k \geq 4n.$$

It is easy to check (by trial and error) that the smallest possible solution is $n = 25$, $k = 20$.

The example shows that at least 2^{20} (about one million) codewords of length 25 are needed to achieve the required rate $\rho = 0.8$. In Section 8.4 we shall give a simple method for constructing a suitable code with these parameters, but of course it is too large to be written down explicitly. For smaller rates, such as $\rho = 0.5$, we can give simpler examples.

Example 7.5

Let C be the code that assigns to each block of three bits, say $y_1y_2y_3$, the codeword $x_1x_2x_3x_4x_5x_6 \in \mathbb{B}^6$ defined as follows:

$$\begin{aligned}
x_1 &= y_1 \\
x_2 &= y_2 \\
x_3 &= y_3 \\
x_4 &= 0 \quad \text{if } y_1 = y_2, \quad x_4 = 1 \quad \text{otherwise,} \\
x_5 &= 0 \quad \text{if } y_2 = y_3, \quad x_5 = 1 \quad \text{otherwise,} \\
x_6 &= 0 \quad \text{if } y_1 = y_3, \quad x_6 = 1 \quad \text{otherwise.}
\end{aligned}$$

Show that C is a 1-error-correcting code, and its rate is $\rho = 0.5$.

Solution Explicitly the code is

$$000 \mapsto 000000, \quad 001 \mapsto 001011, \quad 010 \mapsto 010110, \quad 100 \mapsto 100101,$$

$$011 \mapsto 011101, \quad 101 \mapsto 101110, \quad 110 \mapsto 110011, \quad 111 \mapsto 111000.$$

Checking the distances between pairs of codewords, it turns out that the minimum distance is 3. (In Chapter 8 we shall describe a quicker method of verifying this fact.) Thus C is indeed a 1-error correcting code, and since $n = 6$ and $k = 3$ the information rate is $1/2$.

In the rest of this chapter we shall study the general problem illustrated in the preceding examples. As well as the problem of choosing a suitable code C, we must consider how the situation is affected by changes in the source distribution $\mathbf{p}$. The aim, of course, is to make $M(C, \mathbf{p})$ as small as we please, for any $\mathbf{p}$.

EXERCISES

7.8. Suppose the Sender wishes to code blocks of size k with words of length n, using a 1-error correcting code. If it is required to transmit with information rate not less than 0.6, what are the smallest possible values of n and k?

7.9. In Example 7.5 there is a code for blocks of size 3 using words of length 6. An alternative would be to code each block by repeating it twice: for example, $101 \mapsto 101101$. Compare the parameters of the two codes.

7.10. Is it possible to carry out a construction similar to that in Example 7.5, so that blocks of size *four* are represented by codewords of length *seven*, and the minimum distance is 3? [In Chapter 8 we consider this problem in more general terms.]

7.3 Transmission using the extended BSC

In this section we consider stage T2 of our model, transmission of the encoded stream using an extended BSC, with the possibility that bit-errors may occur. We begin by showing that if the capacity of the BSC Γ is γ, then the capacity of Γ^n is $n\gamma$. Intuitively, this is because Γ^n can be regarded as n copies of Γ 'in parallel', and the copies are independent. The encoded stream may have complex interdependencies among its bits, but when each individual bit is transmitted through Γ^n there is the same probability that an error will occur, irrespective of what happens to the other bits. This fact is implicit in the general definition of Γ^n as the product of n copies of Γ (Definitions 6.6 and 6.7).

In our model the inputs to Γ^n are codewords $c_1 \dots c_n \in C$ and the outputs are words $z_1 \dots z_n \in \mathbb{B}^n$. We have

$$(\Gamma^n)_{c_1 \dots c_n \; z_1 \dots z_n} = \Pr(z_1 \dots z_n \mid c_1 \dots c_n) = \Pr(z_1 \mid c_1) \cdots \Pr(z_n \mid c_n)$$
$$= \Gamma_{c_1 z_1} \cdots \Gamma_{c_n z_n}.$$

For the BSC we have an explicit formula for the entries of Γ^n, but it is not needed here.

As in Section 7.1, we suppose that the encoded stream is emitted by a source $(C, \mathbf{p})$. Then the received stream is a source $(\mathbb{B}^n, \mathbf{q})$, where $\mathbf{q} = \mathbf{p}\Gamma^n$. We shall calculate the capacity of Γ^n by the method described in Section 5.5.

Lemma 7.6

Let Γ be the BSC with bit-error probability e. Then, with the notation as above,

$$H(\mathbf{q} \mid \mathbf{p}) = nh(e).$$

Proof

In Section 5.5 we defined the uncertainty of the output for a given input $c_1 \dots c_n$ as

$$H(\mathbf{q} \mid c_1 \dots c_n) = \sum_{z_1 \dots z_n} (\Gamma^n)_{c_1 \dots c_n \; z_1 \dots z_n} \log \frac{1}{(\Gamma^n)_{c_1 \dots c_n \; z_1 \dots z_n}}.$$

Using the definition of Γ^n, this can be written as

$$\sum_{z_1 \dots z_n} \Gamma_{c_1 z_1} \cdots \Gamma_{c_n z_n} \log \frac{1}{\Gamma_{c_1 z_1} \cdots \Gamma_{c_n z_n}}$$

$$= \sum_{z_1} \cdots \sum_{z_n} \Gamma_{c_1 z_1} \cdots \Gamma_{c_n z_n} \sum_{i=1}^{n} \log \frac{1}{\Gamma_{c_i z_i}}.$$

Now $\sum_z \Gamma_{cz} = 1$ for each c, so the expression reduces to

$$\sum_{i=1}^{n} \sum_{z_i} \Gamma_{c_i z_i} \log \frac{1}{\Gamma_{c_i z_i}}.$$

The values of z_i are 0 and 1, so each sum over z_i is simply

$$\Gamma_{c_i0} \log \frac{1}{\Gamma_{c_i0}} + \Gamma_{c_i1} \log \frac{1}{\Gamma_{c_i1}}.$$

One summand is $e \log(1/e)$ and the other is $(1-e)\log(1/(1-e))$. Hence the sum is $h(e)$, and it follows that

$$H(\mathbf{q} \mid c_1 \dots c_n) = nh(e) \quad \text{for all } c_1 \dots c_n.$$

It follows from Lemma 5.13 that

$$H(\mathbf{q} \mid \mathbf{p}) = \sum_{c_1 \dots c_n} \mathbf{p}(c_1 \dots c_n)\, H(\mathbf{q} \mid c_1 \dots c_n) = nh(e).$$

□

Theorem 7.7

If the capacity of the BSC Γ is $\gamma = 1 - h(e)$, then the capacity of Γ^n is $n\gamma$.

Proof

We use the fact (Section 5.5) that the capacity is the maximum value of

$$H(\mathbf{q}) \;-\; H(\mathbf{q} \mid \mathbf{p}).$$

Lemma 7.6 shows that in this case $H(\mathbf{q} \mid \mathbf{p})$ is a constant, $nh(e)$, so we have only to maximize $H(\mathbf{q})$.

Since $\mathbf{q}$ is a distribution on the set $\mathbb{B}^n$ of size 2^n, it follows from Theorem 3.11 that $H(\mathbf{q}) \leq \log_2(2^n) = n$, and the bound is attained if and only if all elements of $\mathbb{B}^n$ are equally probable. Hence

$$H(\mathbf{q}) - H(\mathbf{q} \mid \mathbf{p}) \;\leq\; n - nh(e) = n\gamma,$$

and the bound is attained. □

We now consider the uncertainty of the situation from the Receiver's viewpoint. In Chapter 5 we denoted this quantity by $H(\Gamma^n; \mathbf{p}) = H(\mathbf{p} \mid \mathbf{q})$. In this case it represents the uncertainty (per codeword) of the encoded stream $(C, \mathbf{p})$, given the received stream. Since each codeword has n bits, the uncertainty per bit is $H(\Gamma^n; \mathbf{p})/n$.

The following theorem shows that, when the required information rate exceeds the capacity of the channel, this quantity cannot be made arbitrarily small for all distributions $\mathbf{p}$, however C and n are chosen.

Theorem 7.8

Let $C \subseteq \mathbb{B}^n$ be a code with information rate ρ, and let $\mathbf{p}^*$ be the distribution in which each codeword in C is equally probable. Suppose that the stream emitted by the source $(C, \mathbf{p}^*)$ is transmitted through the extended BSC Γ^n, where Γ has capacity γ. Then

$$H(\Gamma^n; \mathbf{p}^*) \geq n(\rho - \gamma).$$

Proof

By definition, the capacity of Γ^n is the maximum of $H(\mathbf{p}) - H(\Gamma^n; \mathbf{p})$, taken over all distributions $\mathbf{p}$. In particular, when $\mathbf{p} = \mathbf{p}^*$,

$$H(\mathbf{p}^*) - H(\Gamma^n; \mathbf{p}^*) \leq n\gamma.$$

Now $H(\mathbf{p}^*) = \log |C|$ (Theorem 3.11), so we have

$$H(\Gamma^n; \mathbf{p}^*) \geq \log |C| - n\gamma.$$

Since the information rate of C is $\rho = \log |C|/n$, the inequality can be written in the form stated above. □

This result is trivial when $\rho < \gamma$, since the right-hand side is negative and it is always true (by definition) that the left-hand side is non-negative. However, as we shall see, the result is highly significant when $\rho > \gamma$.

EXERCISES

7.11. Suppose that a binary code with information rate 0.9 is being transmitted through an extended BSC with bit-error probability 0.03. Does the rate exceed the capacity? If each codeword is equally probable, what can be said about the uncertainty from the viewpoint of the Receiver?

7.12. Let Γ be an arbitrary channel. Denote its 2-fold extension by Γ^2, and the input distribution by $\mathbf{p}$, so that the output distribution is $\mathbf{q} = \mathbf{p}\Gamma^2$. Use the method given in Lemma 7.6 to show that

$$H(\mathbf{q} \mid \mathbf{p}) = H(\mathbf{q}' \mid \mathbf{p}') + H(\mathbf{q}'' \mid \mathbf{p}''),$$

where $\mathbf{p}'$ and $\mathbf{p}''$ are the marginal distributions associated with $\mathbf{p}$, and $\mathbf{q}' = \mathbf{p}'\Gamma$, $\mathbf{q}'' = \mathbf{p}''\Gamma$.

7.13. Deduce from the previous result that if the capacity of Γ is γ then the capacity of Γ^2 is 2γ.

7.4 The rate should not exceed the capacity

In this section we shall prove that if the probability of a mistake is required to be arbitrarily small, then the capacity of the channel is an upper bound to the rate at which information can be transmitted.

Consider stage T3 of our model, where the Receiver converts the *received stream* into a *final stream* using the MD rule. The final stream is a sequence of codewords, purporting to have been produced by the source $(C, \mathbf{p})$, and there are two elements of uncertainty associated with it.

The first element of uncertainty arises from the fact that the Receiver does not know whether a mistake has been made. The probability of a mistake is $M = M(C, \mathbf{p})$, and the probability of no mistake is $1 - M$. The uncertainty associated with this situation is $h(M)$, where as usual

$$h(M) = M \log(1/M) + (1 - M) \log(1/(1 - M)).$$

The second element of uncertainty rises from the fact that, if a mistake *has* been made, then the correct codeword has been replaced by an incorrect one. But in this case the Receiver does not know which of the $|C| - 1$ other codewords is the correct one. The probability of a mistake is M, and the uncertainty associated with $|C| - 1$ choices is at most $\log(|C| - 1)$, so this makes a contribution of at most $M \log(|C| - 1)$ to the total uncertainty.

The application of the decision rule can only increase the uncertainty. Thus the uncertainty of the final stream, which is the sum of the two quantities described above, is an upper bound for the uncertainty of the received stream:

$$H(\Gamma^n; \mathbf{p}) \;\leq\; h(M) + M \log(|C| - 1).$$

This result is known as *Fano's inequality*, and it will play an important part in our next theorem. A formal proof is given in Section 7.6.

Recall that the aim is to encode a stream of bits in such a way that

- •1 information is transmitted at a given rate ρ;
- •2 the probability of a mistake, M, is arbitrarily small.

In order to satisfy these criteria we must construct codes $C_n \subseteq \mathbb{B}^n$ for a infinite sequence of values of n. If $|C_n| = 2^{k_n}$, criterion •1 will be satisfied provided that k_n is at least ρn. If that is so, we can split the stream into blocks of size k_n and assign a codeword in C_n to each block. However, the next result shows that we cannot also satisfy •2 if ρ is greater than γ.

Theorem 7.9

Suppose that, for an infinite sequence of values of n, we have constructed codes $C_n \subseteq \mathbb{B}^n$ such that $|C_n| \geq 2^{\rho n}$. Let $\mathbf{p}^*$ be the equiprobable distribution on C_n. If $\rho > \gamma$, then the probability of a mistake $M(C_n, \mathbf{p}^*)$ does not tend to zero as $n \to \infty$.

Proof

Let $|C_n| = 2^{k_n}$ where $k_n \geq \rho n$, and $M_n = M(C_n, \mathbf{p}^*)$. Fano's inequality says that

$$H(\Gamma^n; \mathbf{p}) \;\leq\; h(M_n) + M_n \log(|C_n| - 1).$$

Since $h(M_n) \leq 1$ and $\log(|C_n| - 1) < \log |C_n| = k_n$, it follows that, for any $\mathbf{p}$,

$$H(\Gamma^n; \mathbf{p}) \;<\; 1 + M_n k_n.$$

On the other hand, it follows from the proof of Theorem 7.8 that

$$H(\Gamma^n; \mathbf{p}^*) \geq \log |C_n| - n\gamma = k_n - n\gamma.$$

Combining these inequalities we get

$$1 + M_n k_n > k_n - n\gamma,$$

and so

$$M_n > 1 - \frac{n\gamma + 1}{k_n} \geq 1 - \frac{n\gamma + 1}{n\rho}.$$

As $n \to \infty$ the last expression approaches $1 - \gamma/\rho$, and since $\rho > \gamma$ this is strictly positive. Hence the limit of M_n (if it exists) is not zero. □

EXERCISES

7.14. Suppose that the extended BSC is being used to transmit codewords of length 18, and the MD rule is being used by the Receiver. It is found experimentally that a code with about 64000 codewords can be transmitted with negligible probability of a mistake. What conclusion can be drawn about the bit-error probability e?

7.15. Investigate Fano's inequality in the case when $n = 1$, $C = \{0, 1\}$, and Γ is the BSC with $e = 0.5$. What is special about this case?

7.5 Shannon's theorem

The most striking theoretical result in Information Theory is *Shannon's Theorem*, which was also historically the first result in the area. It is logically equivalent to the converse of Theorem 7.9. Thus it asserts that if $\rho < \gamma$ then it *is* possible to find an infinite sequence of codes C_n such that

$$C_n \subseteq \mathbb{B}^n, \quad |C_n| \geq 2^{\rho n}, \quad \text{and } M(C_n, \mathbf{p}) \to 0 \text{ as } n \to \infty.$$

For example, suppose we wish to transmit a stream of bits, using a device for which the bit-error probability is estimated to be 0.03. It is specified that the probability of a mistake must be less than 10^{-6}. Knowing Shannon's Theorem, we could try to design our system in the following way.

- *Step 1* Choose a specific value of ρ less than γ. Here the capacity of the BSC with $e = 0.03$ is $\gamma = 1 - h(e) = 0.8$ approximately, so a suitable value is $\rho = 0.75$.
- *Step 2* On the basis of Shannon's Theorem, choose a code $C_n \subseteq \mathbb{B}^n$ such that

$$|C_n| = 2^k \text{ (where } k \geq 0.75n), \quad \text{and} \quad M(C_n, \mathbf{p}) < 10^{-6}.$$

- *Step 3* Divide the original stream into blocks of length k and encode the blocks using codewords in C_n.
- *Step 4* Transmit the encoded stream using the given device, and apply the MD rule to the received stream.

In practice, we should soon run into difficulties if we tried to implement this plan. At *Step 2* Shannon's Theorem tells us that a suitable code C_n exists, but it does not tell us how to find it. The same problem occurs in *Step 3*, where we have to set up a rule that assigns a codeword of length n to each block of k bits: in other words, we have to specify an encoding function $\mathbb{B}^k \to \mathbb{B}^n$.

Thus, although Shannon's Theorem is a seminal result, it is not a set of instructions for practitioners. The proof is very ingenious, but it does not form the basis of a method for constructing suitable codes, and for that reason we shall not go into the details. Instead we shall move on (in Chapter 8) to describe some of the simple mathematical techniques that can be used to construct good encoding functions $\mathbb{B}^k \to \mathbb{B}^n$ for many values of $\rho = k/n$. Fano's inequality has shown that the attempt is pointless if $\rho > \gamma$, while Shannon's Theorem guarantees that it is worthwhile if $\rho < \gamma$.

EXERCISES

7.16. Suppose we are using a BSC with known bit-error probability 0.05. How should we go about choosing values of k and n so that Shannon's theorem will guarantee the existence of codes with arbitrarily small values of M?

7.6 Proof of Fano's inequality

The proof of Fano's inequality is elementary, but rather long. We begin with some notation and a couple of lemmas. For each $z \in \mathbb{B}^n$, the codeword $\sigma(z) \in C$ will be denoted by z^*. With this notation, given a codeword $c \in C$ the set $F(c)$ defined in Section 7.1 is the set of z such that $z^* \neq c$. As a temporary notation, let K be the set of pairs (c, z) with this property:

$$K = \{(c, z) \in C \times \mathbb{B}^n \mid c \neq z^*\}.$$

Recall that the joint probability distribution $\mathbf{t}$ on $C \times \mathbb{B}^n$ is such that

$$t_{cz} = p_c \Pr(z \mid c) = p_c \, (\Gamma^n)_{cz}.$$

Lemma 7.10

Let $M = M(C, \mathbf{p})$ be the probability of a mistake when the source $(C, \mathbf{p})$ is transmitted through the extended BSC Γ^n and the MD decision rule is used. Then

$$M = \sum_{(c,z)\in K} t_{cz} \quad \text{and} \quad 1 - M = \sum_z t_{z^*z}.$$

Proof

According to Definition 7.1 the probability of a mistake is

$$M = M(C, \mathbf{p}) = \sum_c p_c M_c = \sum_c p_c \sum_{z \in F(c)} (\Gamma^n)_{cz} = \sum_{(c,z)\in K} t_{cz}.$$

Hence

$$1 - M = \sum_{(z,c)\notin K} t_{cz} = \sum_z \sum_{c=z^*} t_{cz} = \sum_z t_{z^*z}.$$

□

Lemma 7.11

If the source $(C, \mathbf{p})$ is transmitted though the extended BSC Γ^n and the probability of receiving z is q_z, then

$$H(\Gamma^n; \mathbf{p}) = \sum_{(c,z)\in K} t_{cz} \log(q_z/t_{cz}) + \sum_z t_{z^*z} \log(q_z/t_{z^*z}).$$

Proof

By definition, $H(\Gamma^n; \mathbf{p}) = H(\mathbf{t}) - H(\mathbf{q})$, where $\mathbf{q} = \mathbf{p}\Gamma^n$ and

$$H(\mathbf{t}) = \sum_{(c,z)} t_{cz} \log(1/t_{cz}), \quad H(\mathbf{q}) = \sum_z q_z \log(1/q_z).$$

Since $q_z = \sum_c t_{cz}$ it follows that

$$H(\mathbf{t}) - H(\mathbf{q}) = \sum_{(c,z)} t_{cz} \log(q_z/t_{cz}) = \sum_z \sum_c t_{cz} \log(q_z/t_{cz}).$$

For each z, the sum over c contains one term with $c = z^*$. Separating this term from the rest, we have

$$H(\Gamma^n; \mathbf{p}) = \sum_z \sum_{c\neq z^*} t_{cz} \log(q_z/t_{cz}) + \sum_z t_{z^*z} \log(q_z/t_{z^*z}).$$

Writing the double sum as a sum over the set K, the result follows. □

Theorem 7.12 (Fano's inequality)

Let $C \subseteq \mathbb{B}^n$, and let $M = M(C, \mathbf{p})$ be the probability of a mistake when the source $(C, \mathbf{p})$ is transmitted through the extended BSC Γ^n, and the MD decision rule is used. Then

$$H(\Gamma^n; \mathbf{p}) \leq h(M) + M \log(|C| - 1).$$

Proof

Using the expressions for M and $1 - M$ obtained in Lemma 7.10 we can write $h(M)$ as the sum of two terms

$$h(M) = \sum_{(c,z)\in K} t_{cz} \log(1/M) + \sum_z t_{z^*z} \log(1/(1-M)).$$

Lemma 7.11 gives a similar expression for $H(\Gamma^n; \mathbf{p})$. Combining these expressions we have

$$H(\Gamma^n; \mathbf{p}) - h(M) = S_1 + S_2,$$

where

$$S_1 = \sum_{(c,z)\in K} t_{cz} \log(Mq_z/t_{cz}), \qquad S_2 = \sum_z t_{z^*z} \log((1-M)q_z/t_{z^*z}).$$

We shall use the Comparison Theorem 3.10 to prove that

$$S_1 \leq M \log(|C|-1), \qquad S_2 \leq 0.$$

For S_1, it is easy to verify that both $v_{cz} = t_{cz}/M$ and $w_{cz} = q_z/(|C|-1)$ are probability distributions on the set K, so that

$$\begin{aligned} 0 \geq \sum_K v_{cz} \log(w_{cz}/v_{cz}) &= \sum_K (t_{cz}/M) \log(Mq_z/((|C|-1)t_{cz}) \\ &= \sum_K (t_{cz}/M) \log(M/t_{cz}) + \sum_K (t_{cz}/M) \log(q_z/(|C|-1)) \\ &= M^{-1}S_1 + \log(1/(|C|-1)). \end{aligned}$$

Thus $S_1 \leq M \log(|C|-1)$. Similarly, for S_2 both q_z and $u_z = t_{z^*z}/(1-M)$ are probability distributions on the set $\mathbb{B}^n$, so that

$$0 \geq \sum_z u_z \log(q_z/u_z) = \sum_z (t_{z^*z}/(1-M)) \log((1-M)q_z/t_{z^*z}) = (1-M)^{-1}S_2.$$

Thus $S_2 \leq 0$. Putting the two bounds together we have the result. □

Further reading for Chapter 7

Fano's inequality is one of the many contributions to information theory that appear in his book [**7.1**].

Proofs of Shannon's Theorem in the form described in Section 7.5 can be found in Welsh's book [**3.5**], and in many other texts. The theorem can be stated and proved in a much more general way, as described in the books by Ash [**5.1**] and McEliece [**5.2**].

7.1 R.M. Fano. *Transmission of Information: A Statistical Theory of Communications.* MIT Press, Cambridge, Mass. (1961).

8
Linear codes

8.1 Introduction to linear codes

The techniques for constructing useful codes can be extended enormously if we endow the symbols with number-like properties. The 'arithmetic' codes for data compression described in Chapter 4 are an example. Now we shall use 'algebraic' methods in order to construct codes for the purpose of error-correction.

In the case of the binary alphabet $\mathbb{B}$, the symbols 0 and 1 can be 'added' and 'multiplied' according to the rules

$$0+0=0, \quad 0+1=1, \quad 1+0=1, \quad 1+1=0;$$
$$0\times 0=0, \quad 0\times 1=0, \quad 1\times 0=0, \quad 1\times 1=1.$$

In the language of Abstract Algebra, we say that the set $\mathbb{B}$, with these operations, is a *field.* We shall emphasize the algebraic structure by using the notation $\mathbb{F}_2$ for this field.

The most useful method of constructing binary codes depends on the fact that the set ${\mathbb{F}_2}^n$ of words of length n in $\mathbb{F}_2$ is a *vector space.* The rules for vector addition and scalar multiplication are the obvious ones:

$$(x_1x_2\ldots x_n)+(y_1y_2\ldots y_n)=(x_1+y_1\ x_2+y_2\ \ldots\ x_n+y_n);$$
$$0(x_1x_2\ldots x_n)=(00\ldots 0), \quad 1(x_1x_2\ldots x_n)=(x_1x_2\ldots x_n).$$

Recall that a subset C of the vector space ${\mathbb{F}_2}^n$ is a *subspace* if

$$\text{whenever } x,y\in C \text{ it follows that } x+y\in C.$$

N. L. Biggs, *An Introduction to Information Communication and Cryptography,*
DOI: 10.1007/978-1-84800-273-9_8, © Springer-Verlag London Limited 2008

Definition 8.1 (Linear code)

A (binary) *linear code* is a subspace C of the vector space ${\mathbb{F}_2}^n$.

For example, the subset $C = \{000, 110, 011\}$ of ${\mathbb{F}_2}^3$ is not a linear code, because $110 + 011 = 101$, which is not in C.

On the other hand, the repetition code $R_n \subseteq {\mathbb{F}_2}^n$, containing the two words $000\cdots00$ and $111\cdots11$ is a linear code for any n, since $111\cdots11 + 111\cdots11 = 000\cdots00$. Note that every linear code must contain the all-zero word $000\cdots00$, since $x + x = 000\cdots00$ for any $x \in {\mathbb{F}_2}^n$.

We have seen that the construction of good codes involves a trade-off between the *information rate* ρ and the *minimum distance* δ. In the case of linear code we can be more specific about these parameters.

Since a linear code C is a subspace of ${\mathbb{F}_2}^n$ it has a *dimension* k, defined as the size of a minimal spanning set (basis). Every element of C can be expressed uniquely as a linear combination of the basis, and so $|C| = 2^k$ for some k in the range $0 \le k \le n$. It follows that the information rate of C is

$$\rho = \frac{\log_2 |C|}{n} = \frac{k}{n}.$$

Thus it is convenient to use k instead of ρ, and we usually describe a linear code by the parameters (n, k, δ). For codes in general, finding δ is tedious, because it requires the comparison of every pair of codewords, but for a linear code there is a relatively easy way.

Definition 8.2 (Weight)

The *weight* $w(x)$ of a word $x \in {\mathbb{F}_2}^n$ is the number of 1's in x. Equivalently, $w(x) = d(x, 0)$, where 0 denotes the all-zero word $000\cdots00$.

Lemma 8.3

For a linear code, the minimum distance δ is equal to the minimum weight of a nonzero codeword.

Proof

Suppose c_1 and c_2 are codewords such that $c_1 \neq c_2$. A bit in $c_1 + c_2$ is 1 if and only if the corresponding bits in c_1 and c_2 are 0 and 1 in some order – that is, they are different. It follows from the definition of d that

$$d(c_1, c_2) \;=\; w(c_1 + c_2).$$

Since C is linear, $c_1 + c_2$ is also in C. Hence each value of $d(c_1, c_2)$ is the weight of a codeword, and the minimum value is the minimum weight. □

Example 8.4

Find the parameters (n, k, δ) of the following linear codes.

$$D_1 = \{000000, 100000, 010000, 110000\},$$

$$D_2 = \{000000, 111000, 000111, 111111\}.$$

Solution D_1 has words of length $n = 6$ and since there are $4 = 2^2$ codewords the dimension is $k = 2$ (and the rate is $\rho = 1/3$). The weights of the nonzero codewords are is $1, 1, 2$, so $\delta = 1$. With this code, no error correction is possible.

D_2 also has dimension 2 and rate $1/3$, but the minimum distance is $\delta = 3$, and it is a 1-error-correcting code.

The construction of linear codes C with given values of the parameters (n, k, δ) is constrained by the *packing bound*, Theorem 6.20. When $|C| = 2^k$, and $\delta \geq 2r + 1$ (so that C is an r-error-correcting code) the condition is

$$2^k \left(1 + \binom{n}{1} + \binom{n}{2} + \cdots \binom{n}{r}\right) \leq 2^n.$$

Example 8.5

Find an upper bound for the number of codewords in a linear code with word length 12, if a 2-error-correcting code is required.

Solution The packing bound is $2^k(1 + 12 + 66) \leq 4096$, so $2^k \leq 50$ approximately. Since k must be a integer the maximum possible value of 2^k is $2^5 = 32$. (Note that, as yet, we have no means of constructing such a code.)

EXERCISES

8.1. Which of the following subsets of ${\mathbb{F}_2}^3$ is a linear code?

$$C_1 = \{000, 110, 101, 011\}, \; C_2 = \{000, 100, 010, 001\}.$$

8.2. Suppose that we wish to send any one of 128 different messages, and each message is to be represented by a binary codeword of length 10. Is it possible to construct a 1-error-correcting linear code satisfying these conditions?

8.3. Professor MacBrain has decided that each student in his department will be allotted an identity number in the form of a binary word.

(a) If there are 53 students, find the least possible dimension of a linear code for this purpose.

(b) If the code must allow for the correction of one bit-error, find the least possible length of the codewords.

8.4. Suppose it is required to construct a linear code with $n = 12$ and $\delta = 3$. Find an upper bound for the information rate of such a code.

8.5. Given words $x, y \in \mathbb{F}_2{}^n$ let $x * y$ denote the word that has 1 in any position if and only if both x and y have 1 in that position. Prove that

$$w(x + y) = w(x) + w(y) - 2w(x * y).$$

Deduce that the set of all words with even weight in $\mathbb{F}_2{}^n$ is a linear code.

8.6. Let $B(n, \delta)$ denote the maximum dimension of a *linear* code in $\mathbb{F}_2{}^n$ with minimum distance δ. Use the packing bound to show that

$$B(6,3) \leq 3, \quad B(7,3) \leq 4, \quad B(8,3) \leq 4.$$

8.2 Construction of linear codes using matrices

In Chapter 7 we studied the problem of transmitting information at a given rate, while reducing the probability of a mistake. The key idea was to split the original stream of bits into blocks of size k and assign a codeword of length n to each of the 2^k possible blocks. In other words, we require an encoding function $\mathbb{F}_2{}^k \to \mathbb{F}_2{}^n$. Given that $\mathbb{F}_2{}^k$ and $\mathbb{F}_2{}^n$ are vector spaces, an obvious candidate for such a function is a *linear transformation*, defined by a matrix.

For the purposes of matrix algebra it is often convenient to replace a word x, regarded as a row vector, by the corresponding *column vector* x' (the transpose of x). Thus, if E is a $n \times k$ matrix over $\mathbb{F}_2$ and y is in $\mathbb{F}_2{}^k$ then then the word x defined by $x' = Ey'$ is in $\mathbb{F}_2{}^n$. The matrix E defines an encoding function $\mathbb{F}_2{}^k \to \mathbb{F}_2{}^n$.

Example 8.6

In Example 7.5 we gave a rule that assigns to each block $y_1y_2y_3$ of size 3 a codeword $x_1x_2x_3x_4x_5x_6$ of length 6. Express this rule in matrix form.

Solution The rule for x_4 (for example) is that $x_4 = 0$ if $y_1 = y_2$ and $x_4 = 1$ otherwise. Using the algebraic structure of the field $\mathbb{F}_2$ this can be expressed by the linear equation $x_4 = y_1 + y_2$. In the same way, we have the equations $x_5 = y_2 + y_3$, $x_6 = y_1 + y_3$. All the equations can be written as a single matrix equation:

$$\begin{pmatrix} x_1 \\ x_2 \\ x_3 \\ x_4 \\ x_5 \\ x_6 \end{pmatrix} = \begin{pmatrix} y_1 \\ y_2 \\ y_3 \\ y_1 + y_2 \\ y_2 + y_3 \\ y_1 + y_3 \end{pmatrix} = E \begin{pmatrix} y_1 \\ y_2 \\ y_3 \end{pmatrix},$$

where E is the 6×3 matrix given by

$$E = \begin{pmatrix} 1 & 0 & 0 \\ 0 & 1 & 0 \\ 0 & 0 & 1 \\ 1 & 1 & 0 \\ 0 & 1 & 1 \\ 1 & 0 & 1 \end{pmatrix}.$$

As in Example 7.5, we can check that the resulting code $C \subseteq {\mathbb{F}_2}^6$ has parameters $n = 6$, $k = 3$, and $\delta = 3$, and hence it is a 1-error-correcting code with rate $1/2$.

In general, given any $n \times k$ matrix E over $\mathbb{F}_2$, let C be the set

$$\{x \in {\mathbb{F}_2}^n \mid x' = Ey' \text{ for some } y \in {\mathbb{F}_2}^k\}.$$

If x and w are in C, say $x' = Ey'$ and $w' = Ez'$, then $(x + w)' = E(y + z)'$, so $x + w$ is also in C. Thus C is a linear code. Technically, it is the *image* of E.

Essentially, we now have the answer to the practical problem of encoding a stream of bits. The stream is split into blocks y of length k, and the codeword for y is x, where $x' = Ey'$, for a suitable matrix E. Of course, we must try to ensure that resulting code has good error-correction properties, and that there is a workable method of implementing the MD rule. For these purposes a slight modification of the construction is useful, as we shall see in the next section.

EXERCISES

8.7. Suppose we encode a block of bits $y_1 y_2 \dots y_k$ by setting $x_i = y_i$ for $i = 1, 2, \dots k$, and defining the *parity check* bit x_{k+1} as follows:

$$x_{k+1} = \begin{cases} 0 & \text{if an even number of the } y_i\text{'s are 1;} \\ 1 & \text{if an odd number of the } y_i\text{'s are 1.} \end{cases}$$

Write down the matrix E such that $x' = Ey'$.

8.8. Show that in the code defined in previous exercise every word has even weight.

8.9. Find the information rate ρ and the minimum distance δ for the parity check code described in Exercise 8.7.

8.3 The check matrix of a linear code

In the previous section we described a method of coding a stream of bits by dividing it into blocks of length k and applying an $n \times k$ matrix E to each block y. In Example 8.6 the transformation $x' = Ey'$ was defined by the linear equations

$$x_1 = y_1, \qquad x_2 = y_2, \qquad x_3 = y_3,$$

$$x_4 = y_1 + y_2, \qquad x_5 = y_2 + y_3, \qquad x_6 = y_1 + y_3.$$

The variables y_1, y_2, y_3 can be eliminated from the equations, giving

$$x_1 + x_2 + x_4 = 0, \qquad x_2 + x_3 + x_5, \qquad x_1 + x_3 + x_6 = 0.$$

(Remember that $-1 = 1$ in the field $\mathbb{F}_2$.) These equations can be written in the form of a matrix equation $Hx' = 0'$, where

$$H = \begin{pmatrix} 1 & 1 & 0 & 1 & 0 & 0 \\ 0 & 1 & 1 & 0 & 1 & 0 \\ 1 & 0 & 1 & 0 & 0 & 1 \end{pmatrix}.$$

Hence we have an alternative means of defining the code, as the set of x such that $Hx' = 0'$. In Linear Algebra this set is known as the *kernel* or *null space* of H; it is clearly a subspace of ${\mathbb{F}_2}^n$, so we have a linear code. The general situation is as follows.

Lemma 8.7

Let E be an $n \times k$ matrix over $\mathbb{F}_2$, of the form

$$\begin{pmatrix} I \\ A \end{pmatrix},$$

where I is the identity matrix with size k, and A is any $(n-k) \times k$ matrix. Then the code

$$\{x \in {\mathbb{F}_2}^n \mid x' = Ey' \text{ for some } y \in {\mathbb{F}_2}^k\}$$

can also be defined as

$$\{x \in {\mathbb{F}_2}^n \mid Hx' = 0'\},$$

where H is the $(n-k) \times n$ matrix $(A\ I)$. (Here I is the identity matrix with size $n-k$.)

Proof

When E has the stated form the condition $x' = Ey'$ is

$$x' = \begin{pmatrix} y' \\ Ay' \end{pmatrix}, \quad \text{so} \quad Hx' = \begin{pmatrix} A & I \end{pmatrix} \begin{pmatrix} y' \\ Ay' \end{pmatrix} = Ay' + Ay'.$$

This reduces to $0'$ because $0+0$ and $1+1$ are both 0 in $\mathbb{F}_2$.

Conversely, when H has the stated form and $Hx' = 0'$, let $x = (a\ b)$ so that

$$0' = Hx' = \begin{pmatrix} A & I \end{pmatrix} \begin{pmatrix} a' \\ b' \end{pmatrix} = Aa' + b'.$$

Thus $Aa' = b'$, and x is in the image of E since

$$Ea' = \begin{pmatrix} I \\ A \end{pmatrix} a' = \begin{pmatrix} a' \\ Aa' \end{pmatrix} = x'.$$

□

Definition 8.8 (Check matrix)

A matrix H over $\mathbb{F}_2$ with m rows and n columns is the *check matrix* for the linear code

$$C = \{x \in {\mathbb{F}_2}^n \mid Hx' = 0'\}.$$

In the definition H can be any $m \times n$ matrix over $\mathbb{F}_2$. Suppose, as in Lemma 8.7, that H has the *standard form* $(A\ I)$, where A has $n - m$ columns and I has m columns. Then the corresponding code has dimension $k = n - m$. To prove this, let

$$H = (A\ I) = \begin{pmatrix} a_{11} & a_{12} & \cdots & a_{1k} & 1 & 0 & \cdots & 0 \\ a_{21} & a_{22} & \cdots & a_{2k} & 0 & 1 & \cdots & 0 \\ . & . & \cdots & . & . & . & \cdots & . \\ . & . & \cdots & . & . & . & \cdots & . \\ a_{m1} & a_{m2} & \cdots & a_{mk} & 0 & 0 & \cdots & 1 \end{pmatrix}.$$

Then the equations $Hx' = 0'$ are

$$a_{11}x_1 + a_{12}x_2 + \cdots + a_{1k}x_k + x_{k+1} = 0$$

$$a_{21}x_1 + a_{22}x_2 + \cdots + a_{2k}x_k + x_{k+2} = 0$$

$$\cdots \qquad \cdots$$

$$a_{m1}x_1 + a_{m2}x_2 + \cdots + a_{mk}x_k + x_n = 0.$$

These equations can be rearranged so that the values $x_{k+1}, x_{k+2}, \dots x_n$ are defined in terms of the values $x_1, x_2, \dots, x_k$:

$$x_{k+1} = a_{11}x_1 + a_{12}x_2 + \cdots + a_{1k}x_k$$

$$x_{k+2} = a_{21}x_1 + a_{22}x_2 + \cdots + a_{2k}x_k$$

$$\cdots \qquad \cdots$$

$$x_n = a_{m1}x_1 + a_{m2}x_2 + \cdots + a_{mk}x_k.$$

In a typical codeword $x = x_1x_2 \dots x_n$, we say that

$x_1, x_2, \dots, x_k$ are *message bits*

$x_{k+1}, x_{k+2}, \dots, x_n$ are *check bits.*

If the message bits are given arbitrary values, the values of the check bits are determined by the equations displayed above. Since there are 2^k possible values of the message bits, the dimension of the code is k.

Example 8.9

Make a list of the codewords defined by the check matrix

$$H = \begin{pmatrix} 1 & 0 & 1 & 0 \\ 1 & 1 & 0 & 1 \end{pmatrix}.$$

What are the parameters of this code?

Solution If $x = x_1x_2x_3x_4$ then the condition $Hx' = 0'$ means that

$$x_1 + x_3 = 0, \qquad x_1 + x_2 + x_4 = 0.$$

We can rewrite these equations so that x_3 and x_4 are given in terms of x_1 and x_2:

$$x_3 = x_1, \qquad x_4 = x_1 + x_2.$$

The codewords can be found by giving all possible values to x_1, x_2 and using the equations to calculate x_3, x_4.

x_1	x_2	x_3	x_4	weight
0	0	0	0	0
0	1	0	1	2
1	0	1	1	3
1	1	1	0	3

The code has parameters $n = 4$, $k = 2$, and the minimum distance is equal to the minimum nonzero weight, which is $\delta = 2$.

EXERCISES

8.10. Make a list of the codewords defined by the check matrix

$$H = \begin{pmatrix} 1 & 0 & 1 & 0 & 0 \\ 1 & 1 & 0 & 1 & 0 \\ 0 & 1 & 0 & 0 & 1 \end{pmatrix}.$$

What are the parameters of this code?

8.11. Write down a check matrix for the parity check code described in Exercise 8.7.

8.12. The code $C_1 = \{00, 01, 10, 11\}$ may be used to represent four messages $\{a, b, c, d\}$. The code C_r is obtained by repeating each word in C_1 r times: for example, in C_3 the message b is represented by 010101. Show that C_r is a linear code, find its parameters (n, k, δ) and write down a check matrix for it.

8.4 Constructing 1-error-correcting codes

The following theorem describes a very simple way of constructing a check matrix for a 1-error-correcting code.

Theorem 8.10

The code C defined by a check matrix H is a 1-error-correcting code provided that

- •1 no column of H consists entirely of 0's; and
- •2 no two columns of H are the same.

Proof

Suppose there is a codeword $c \in C$ with weight 1, say

$$c = 0 \cdots 010 \cdots 0,$$

where the ith bit is 1. Then, according to the rule for matrix multiplication (Figure 8.1), Hc' is equal to the ith column of H.

$$Hc' = \begin{pmatrix} \cdots & * & \cdots \\ \cdots & * & \cdots \\ & & \\ \cdots & * & \cdots \\ \cdots & * & \cdots \end{pmatrix} \begin{pmatrix} 0 \\ \cdot \\ 0 \\ 1 \\ 0 \\ \cdot \\ 0 \end{pmatrix} = \begin{pmatrix} * \\ * \\ \\ * \\ * \end{pmatrix}.$$

Figure 8.1 Calculation of Hc' when $c = 0 \cdots 010 \cdots 0$

But if c is a codeword $Hc' = 0'$, so the ith column must be the zero column $0'$, contradicting condition •1.

Suppose there is a codeword of weight 2, say

$$c = 0 \cdots 010 \cdots 010 \cdots 0,$$

where the ith and jth bits are 1. Then, by a similar calculation (Figure 8.2), Hc' is equal to the sum of the ith and jth columns of H.

But if c is a codeword, then $Hc' = 0'$. Since we are using arithmetic in the field $\mathbb{F}_2$, this means that the ith column and the jth column are equal, contradicting condition •2. Hence, if the conditions are satisfied, any nonzero codeword must have weight at least 3, and we have a 1-error correcting code. □

$$Hc' = \begin{pmatrix} \cdots & * & \cdots & \dagger & \cdots \\ \cdots & * & \cdots & \dagger & \cdots \\ & & & & \\ \cdots & * & \cdots & \dagger & \cdots \\ \cdots & * & \cdots & \dagger & \cdots \end{pmatrix} \begin{pmatrix} 0 \\ \cdot \\ 0 \\ 1 \\ 0 \\ \cdot \\ 0 \\ 1 \\ 0 \\ \cdot \\ 0 \end{pmatrix} = \begin{pmatrix} * \\ * \\ \\ * \\ * \end{pmatrix} + \begin{pmatrix} \dagger \\ \dagger \\ \\ \dagger \\ \dagger \end{pmatrix}.$$

Figure 8.2 Calculation of Hc' when $c = 00\cdots010\cdots010\cdots0$.

Example 8.11

In Example 7.4 we showed that the smallest possible values of n and k for a 1-error-correcting code with information rate 0.8 are $n = 25$, $k = 20$. How can such a code be constructed using a check matrix?

Solution We require H to be an $m \times n$ matrix with $m = n - k = 5$ rows. According to Theorem 8.10, the $n = 25$ columns must be distinct, and not the zero column. In fact there are $2^5 - 1 = 31$ possible columns and any 25 of them can be chosen. If H is in standard form the last five columns will be the five words of weight 1, and the first 20 columns can be any other nonzero words.

The matrix algebra illustrated in Figure 8.1 provides a remarkably simple way of implementing the MD rule in some circumstances.

Theorem 8.12

Let $C \subseteq \mathbb{F}_2{}^n$ be a linear code defined by a check matrix H. Suppose that one bit-error is made in transmitting a codeword, the received word being z. Then the error has occurred in the ith bit of z, where i is determined by the fact that Hz' is equal to the ith column of H.

Proof

Suppose that the codeword sent is c and the error is made in the ith bit. Then the received word is $z = c + w$, where w is the word $0\cdots010\cdots0$ with 1 in the ith place. Since c is a codeword, $Hc' = 0'$, and (as in Figure 8.1) Hw' is the

ith column of H. Hence $Hz' = H(c+w)' = Hc' + Hw'$ is the ith column of H. □

Assuming that not more than one bit-error is made in transmitting each codeword, the procedure shown in Figure 8.3 can be used.

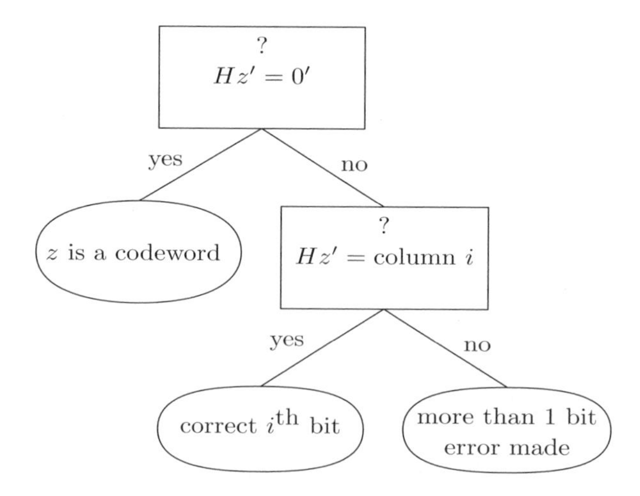

Figure 8.3 Processing a received word z for a code with check matrix H

For each received word z the Receiver should calculate Hz'. If $Hz' = 0$, z is itself the intended codeword. If Hz' is equal to a column of H, say the ith column, then the intended codeword is obtained by changing the ith bit in z. In any other case, at least two bit-errors have occurred.

EXERCISES

8.13. Construct a 1-error-correcting code $C \in \mathbb{F}_2{}^6$ with $|C| = 8$.

8.14. Find the smallest values of n and k for which a 1-error correcting code with information rate 0.75 can exist, and explain how to define such a code by means of a check matrix.

8.15. Write down all the codewords belonging to the linear code with the following check matrix, and find the parameters (n, k, δ) for this code.

$$\begin{pmatrix} 1 & 1 & 0 & 1 & 0 & 0 & 1 \\ 0 & 0 & 0 & 1 & 1 & 0 & 1 \\ 1 & 0 & 1 & 1 & 0 & 0 & 1 \\ 0 & 0 & 0 & 0 & 0 & 1 & 1 \end{pmatrix}.$$

8.16. In Exercise 8.2 we showed that it is impossible to encode 128 different messages in such a way that one error is corrected, using binary words of length 10. Explain how to construct a linear code satisfying the conditions, using words of length 11.

8.17. A codeword from the code defined by the following check matrix is sent, and the word 111010 is received.

$$\begin{pmatrix} 1 & 1 & 1 & 1 & 0 & 0 \\ 1 & 1 & 0 & 0 & 1 & 0 \\ 1 & 0 & 1 & 0 & 0 & 1 \end{pmatrix}.$$

What was the intended codeword, assuming that only one error has been made?

8.18. In the preceding exercise, how many received words cannot be corrected on the assumption that at most one error has been made?

8.5 The decoding problem

In general, there is no easy way of implementing the MD decision rule. Given a set of codewords $C \subseteq \mathbb{B}^n$ and a word $z \in \mathbb{B}^n$ the problem is to find a codeword $c \in C$ that is nearest to z, in the sense of Hamming distance. Of course, it can be done by the brute-force method of running through a list of all the codewords, but that is not an efficient method.

When C is defined by means of an algebraic construction, there may be ways of improving on the brute-force approach. For example, if the code is defined by a check matrix and we can assume that only one bit-error has occurred, the simple rule stated in Theorem 8.12 can be used. More generally, the decoding problem can often be simplified by using a a technique known as *syndrome decoding*.

Definition 8.13 (Syndrome)

Let C be the linear code defined by a check matrix H. For any word $z \in \mathbb{F}_2{}^n$ the *syndrome* of z is s, where $Hz' = s'$.

If z is a codeword then the syndrome of z is the zero word. If z is the result of making a single error, in the ith bit of a codeword, then the syndrome is the word in the ith column of H.

Lemma 8.14

Two words $y, z \in \mathbb{F}_2{}^n$ have the same syndrome with respect to C if and only if $y = z + c$, for some $c \in C$.

Proof

It follows from the definitions that

$$Hy' = Hz' \iff H(y' - z') = 0' \iff y - z \in C \iff y = z + c.$$

$\square$

The set of $y \in \mathbb{F}_2{}^n$ such that $y = z + c$ for some $c \in C$ is known as the *coset* of z with respect to C, and is denoted by $z + C$. We recall from elementary Group Theory that two cosets are either disjoint or identical (they cannot partly overlap), so the distinct cosets form a partition of $\mathbb{F}_2{}^n$.

Example 8.15

List the cosets of the code $C \subseteq \mathbb{F}_2{}^4$ defined by the check matrix

$$H = \begin{pmatrix} 1 & 0 & 1 & 0 \\ 1 & 1 & 0 & 1 \end{pmatrix}.$$

Solution In Example 8.9 we showed that C contains four codewords, 0000, 0101, 1011, 1110. Since $|C| = 4$ and $|\mathbb{F}_2{}^4| = 2^4 = 16$, there are $16/4 = 4$ distinct cosets. The coset $0000 + C$ is C itself. A distinct coset $z + C$ can be constructed by choosing any z that is not in this coset, such as $z = 1000$. Continuing in this way, we obtain the four distinct cosets, as listed below.

$0000 + C$	$1000 + C$	$0100 + C$	$0010 + C$
0000	1000	0100	0010
0101	1101	0001	0111
1011	0011	1111	1001
1110	0110	1010	1100

If the code has word-length n and dimension k, the number of cosets is $2^n/2^k = 2^{n-k} = 2^m$, where m is the number of rows of H. Since each syndrome is a word of length m, and different cosets have different syndromes, all possible words in $\mathbb{F}_2{}^m$ occur as syndromes. It is convenient to arrange them in a definite order, such as the dictionary order (Definition 4.15).

Definition 8.16 (Coset leader, syndrome look-up table)

A *coset leader* is a word of least weight in its coset; if there are several possibilities, we choose one for definiteness. A *syndrome look-up table* is an ordered list of pairs (s, f) such that, for each syndrome s, f is the coset leader for the coset that has syndrome s.

In the list of cosets given in Example 8.15, the first element of each coset has least weight, and can be chosen as the coset leader. However, in the third coset there are two words of weight 1, and either of them could be chosen. If we choose 0100 from this coset, and arrange the syndromes in dictionary order, the syndrome look-up table is as follows.

syndrome :	00	01	10	11
coset leader :	0000	0100	0010	1000

The syndrome decoding method is based on the assumption that the Receiver knows the check matrix H for a code C and has a copy of the look-up table. In that case, the following decision rule $\sigma : \mathbb{F}_2{}^n \to C$ can be applied.

- For a given received word z calculate the syndrome s, using $s' = Hz'$.
- Look up the coset leader f corresponding to s.
- Define $\sigma(z) = z + f$.

Theorem 8.17

With the notation used above,

(i) $\sigma(z) = z + f$ is a codeword;
(ii) there is no codeword $c \in C$ such that $d(z, c) < d(z, z + f)$.

Proof

(i) Since f and z have the same syndrome s, they are in the same coset, and $f = z + c^*$ for some $c^* \in C$. Thus

$$z + f = z + (z + c^*) = (z + z) + c^* = c^*, \quad \text{which is in } C.$$

(ii) We have

$$d(z, z+f) = w(z+z+f) = w(f).$$

For any codeword c, $d(z,c) = w(z+c)$, so if $d(z,c) < d(z, z+f)$ it must follow that $w(z+c) < w(f)$. But $z+c$ and z are in the same coset, so $z+c$ and f are in the same coset, and since f is a coset leader, $w(z+c) \geq w(f)$. □

The first part of the theorem shows that σ is a valid decision rule. The second part shows that it is an MD rule. The choice of coset leader corresponds to choosing one of the codewords nearest to z as the codeword $\sigma(z)$.

Example 8.18

The check matrix

$$H = \begin{pmatrix} 1 & 0 & 1 & 0 & 0 \\ 1 & 1 & 0 & 1 & 0 \\ 0 & 1 & 0 & 0 & 1 \end{pmatrix}$$

defines the code $C = \{00000, 01011, 10110, 11101\}$. Construct a syndrome look-up table for C and use it to determine $\sigma(10111)$.

Solution The cosets can be constructed in the usual way. The first coset is $00000 + C = C$. This does not contain 10000 (for example), so the next coset is $10000 + C$. Continuing in this way we obtain eight cosets

$$00000+C \quad 10000+C \quad 01000+C \quad 00100+C$$

$$00010+C \quad 00001+C \quad 00101+C \quad 00111+C.$$

Note that the representatives used to construct the cosets are not necessarily the coset leaders. To find the coset leaders we must choose one word from each coset that has minimum weight. For example, the coset $00111+C$ contains the words 00111, 01100 10001, and 11010. Thus the coset leader could be either 01100 or 10001.

Suppose we choose 01100 as the coset leader for this coset, and make similar choices for the other cosets. For each coset leader f, the syndrome s for all words in that coset is given by $s' = Hf'$. The following is a syndrome look-up table for C (other choices for the coset leaders are possible).

000	001	010	011	100	101	110	111
00000	00001	00010	01000	00100	00101	10000	01100

Suppose $z = 10111$ is the received word. Since $Hz' = 001'$ the corresponding coset leader is $f = 00001$ and the Receiver will decide that

$$\sigma(z) = z + f = 10110.$$

It should not be thought that syndrome decoding, in itself, is always an improvement on the brute-force method of implementing the MD rule. If the received word z is compared with each codeword in order to find the nearest one, then, for a linear code of dimension k, 2^k comparisons are needed. With syndrome decoding the Receiver must have a copy of the look-up table, which contains 2^{n-k} entries, and this too may be a large number, so the method cannot be regarded as 'efficient' in a strict sense. Fortunately, when specific codes are constructed by algebraic methods (as in the next chapter) there may be better ways of implementing the rule.

EXERCISES

8.19. Construct a syndrome look-up table for the code defined by the check matrix

$$\begin{pmatrix} 1 & 0 & 1 & 0 & 0 \\ 0 & 1 & 0 & 1 & 0 \\ 1 & 1 & 0 & 0 & 1 \end{pmatrix}.$$

Use your table to determine the codewords corresponding to the following received words:

$$11111, \quad 11010, \quad 01101, \quad 01110.$$

8.20. Suppose the Sender uses the code

$$C = \{00000, 01011, 10110, 11101\}$$

and the Receiver uses the syndrome look-up table for C given in Example 8.18. Given that the probability of a bit-error in transmission is e, estimate $M_{\mathbf{0}}$, the probability of a mistake when the codeword 00000 is sent.

8.21. Denote by $\mathbf{h}_i$ the column 4-vector corresponding to the binary representation of the integer i $(1 \leq i \leq 15)$. For example, $\mathbf{h}_9 = [1001]'$. Let H be the 8×15 binary matrix with the first four rows

$$\mathbf{h}_1\ \mathbf{h}_2\ \mathbf{h}_3\ \mathbf{h}_4\ \mathbf{h}_5\ \mathbf{h}_6\ \mathbf{h}_7\ \mathbf{h}_8\ \mathbf{h}_9\ \mathbf{h}_{10}\ \mathbf{h}_{11}\ \mathbf{h}_{12}\ \mathbf{h}_{13}\ \mathbf{h}_{14}\ \mathbf{h}_{15},$$

and the last four rows

$$\mathbf{h}_5\ \mathbf{h}_3\ \mathbf{h}_1\ \mathbf{h}_1\ \mathbf{h}_8\ \mathbf{h}_{15}\ \mathbf{h}_1\ \mathbf{h}_{15}\ \mathbf{h}_8\ \mathbf{h}_5\ \mathbf{h}_5\ \mathbf{h}_8\ \mathbf{h}_3\ \mathbf{h}_{15}\ \mathbf{h}_3.$$

H is the check matrix for a code $C \subseteq \mathbb{F}_2{}^{15}$. A codeword $c \in C$ is transmitted and the received word is

$$z = 100\,100\,100\,100\,000.$$

Calculate the syndrome of z. How do we know that z is not a codeword, and that more than one bit-error has been made? Assuming that exactly two bit-errors have been made, which two bits should be corrected? [The general construction of which this is an example will be discussed in the next chapter.]

Further reading for Chapter 8

The standard texts by MacWilliams and Sloane [**6.2**], and Pless and Huffman [**6.3**] are recommended for background reading on error-correcting codes.

9
Algebraic coding theory

9.1 Hamming codes

In the previous chapter we showed that algebraic methods can be used to construct useful codes. To set the scene for further developments we begin by describing an important family of codes, discovered by R.W. Hamming in 1950.

According to Theorem 8.10 a binary linear code with $\delta \geq 3$ can be defined by a check matrix in which all the columns are distinct and non-zero. If the number m of rows is given, then there are 2^m column vectors of length m, and so the maximum number of distinct non-zero columns is $2^m - 1$.

Definition 9.1 (Hamming code)

A code defined by a binary check matrix with m rows and $2^m - 1$ distinct non-zero columns is a binary *Hamming code*.

Note that the order in which the columns are listed is unimportant, although in practice a definite order is obviously needed. Rearranging the rows just involves a corresponding rearrangement of the bits in each codeword, and does not affect the parameters of the code. Two codes that are related in this way are said to be *equivalent*.

N. L. Biggs, *An Introduction to Information Communication and Cryptography*,
DOI: 10.1007/978-1-84800-273-9_9,

Example 9.2

Find the check matrix and the parameters of the Hamming code with $m = 3$.

Solution There are $2^3 - 1 = 7$ columns. One way of listing them would be to put them in the *dictionary order* (Definition 4.15). But, as we shall see shortly, there are good reasons for using the *numerical order*, which is the order corresponding to the binary representations of the numbers $1, 2, 3, 4, 5, 6, 7$:

$$H_1 = \begin{pmatrix} 0 & 0 & 0 & 1 & 1 & 1 & 1 \\ 0 & 1 & 1 & 0 & 0 & 1 & 1 \\ 1 & 0 & 1 & 0 & 1 & 0 & 1 \end{pmatrix}.$$

Alternatively, we could arrange the columns so that those with weight 1 come last, obtaining a check matrix in the *standard form*:

$$H_2 = \begin{pmatrix} 0 & 1 & 1 & 1 & 1 & 0 & 0 \\ 1 & 0 & 1 & 1 & 0 & 1 & 0 \\ 1 & 1 & 0 & 1 & 0 & 0 & 1 \end{pmatrix}.$$

Clearly, the codes defined by the matrices H_1 and H_2 are equivalent.

The parameters (n, k, δ) are easily found. The word-length is $n = 7$, and the dimension is $k = n - m = 4$. The minimum distance δ is at least 3, since all the columns of H_1 are distinct and non-zero, and it is easy to check that 1110000 is a codeword with weight 3, so δ is exactly 3.

Generally, for each $m \geq 3$, there is a binary Hamming code with parameters

$$n = 2^m - 1, \quad k = 2^m - 1 - m, \quad \delta = 3.$$

These codes have a very special property.

Theorem 9.3

In a binary Hamming code, every word in ${\mathbb{F}_2}^n$ is either a codeword or is at distance 1 from exactly one codeword.

Proof

There are 2^k codewords, and for each $c \in C$ the size of the neighbourhood $N_1(c)$ is $1 + n$. We know that $\delta = 3$, so these neighbourhoods are disjoint. Substituting the appropriate values of the parameters, we have

$$2^k(1 + n) = 2^k \times 2^m = 2^{m+k} = 2^{2^m - 1} = 2^n.$$

This means that the 2^k codewords $c \in C$ determine disjoint neighbourhoods $N_1(c)$ that completely cover ${\mathbb{F}_2}^n$, as claimed. $\square$

Generally, an r-error correcting code has the property that $\delta \geq 2r+1$, so the neighbourhoods $N_r(c)$ $(c \in C)$ are mutually disjoint. However, there will usually be 'gaps' not covered by these neighbourhoods, so that some words are not in any of the neighbourhoods $N_r(c)$.

Definition 9.4 (Perfect code)

An r-error-correcting code C is said to be a *perfect code* if every word is in one of the neighbourhoods $N_r(c)$ for some $c \in C$.

Theorem 9.3 asserts that a binary Hamming code is a perfect code with $r = 1$. In the case $m = 3$ there are $2^4 = 16$ codewords c, and each neighbourhood $N_1(c)$ contains $1 + 7 = 8$ words. These 16 neighbourhoods are disjoint and, since $16 \times 8 = 128$ is the total number of words, the code is perfect (see Figure 9.1).

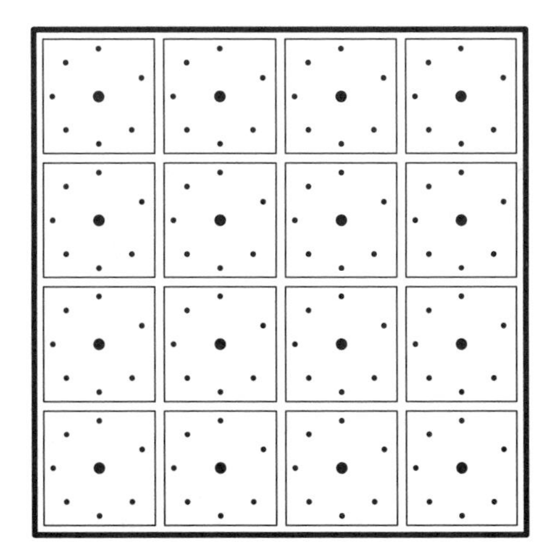

Figure 9.1 The Hamming code with parameters $(7, 4, 3)$ is perfect

The main reason for using algebraic methods is that the algebraic structure can provide good algorithms for encoding and decoding. For the binary Hamming codes, we can use the following techniques.

• *Encoding* Suppose the columns of the check matrix H are arranged in numerical order. Then the bits $x_1, x_2, x_4, \ldots, x_{2^{m-1}}$ are check bits, and we can write down the equations that define them in terms of the remaining (message)

bits. For example, in the case $m = 3$ the equations derived from H_1 are

$$x_1 = x_3 + x_5 + x_7, \quad x_2 = x_3 + x_6 + x_7, \quad x_4 = x_5 + x_6 + x_7.$$

Sometimes it is convenient to use the first k bits as the message bits and the remaining $n - k$ bits as the check bits. This corresponds to writing the check matrix in the standard form H_2, whose columns are obtained from those of H_1 by the permutation

$$1 \mapsto 7,\ 2 \mapsto 6,\ 3 \mapsto 1,\ 4 \mapsto 5,\ 5 \mapsto 2,\ 6 \mapsto 3,\ 7 \mapsto 4.$$

Applying this permutation, and reordering the equations, the check bits are given by

$$x_5 = x_2 + x_3 + x_4, \quad x_6 = x_1 + x_3 + x_4, \quad x_7 = x_1 + x_2 + x_4.$$

• *Decoding* When the columns of the check matrix are arranged in numerical order, the syndrome decoding rule takes a very simple form. First, we observe (Exercise 9.3) that there are $n + 1$ cosets

$$0 + C, \quad e_1 + C, \quad e_2 + C, \quad \ldots, \quad e_n + C,$$

where e_i denotes the word $0 \cdots 010 \cdots 0$ in which the ith bit is 1. The syndrome of the coset $e_i + C$ is He_i', which is equal to the ith column of H. But this is just the binary representation of i. So the rule for correcting a single error is: if the syndrome of the received word z is the binary representation of i, then the ith bit is in error.

EXERCISES

9.1. Write down the check matrix for the Hamming code of length 15, using the numerical order of the columns. How many codewords are there? Which of the following words are codewords?

$$011010110110000, \qquad 100000100000011, \qquad 110010110111111.$$

Correct those that are not codewords, assuming that only one error has been made.

9.2. Show that if the check matrix for a Hamming code is arranged so that the columns are in numerical order, then the word $11100 \cdots 00$ is a codeword.

9.3. Prove that the number of cosets of the the binary Hamming code C with words of length n is $n + 1$, and show that a complete set of coset representatives is the set of $n + 1$ words $0, e_1, e_2, \ldots e_n$, as defined above.

9.2 Cyclic codes

The Hamming codes provide a family of examples in which the dimension k of the of the code, and thus its size $|C| = 2^k$, can be as large as we please, while the minimum distance is constant.

Question: Is it possible to construct explicitly families of linear codes in which both the dimension and the minimum distance can be as large as we please?

In order to answer this question positively we shall give (i) a general method for constructing linear codes, based on simple algebraic ideas, and (ii) a specific construction using the general method.

In the rest of this chapter we shall use the notation $\mathbf{a} = a_0a_1 \dots a_{n-1}$ to represent a typical word in $\mathbb{F}_2{}^n$. The *cyclic shift* of $\mathbf{a}$ is the word

$$\hat{\mathbf{a}} = a_{n-1}a_0a_1 \dots a_{n-2}.$$

Definition 9.5 (Cyclic code)

A set $C \subseteq \mathbb{F}_2{}^n$ is said to be a *cyclic code* if

(i) it is a linear code, and

(ii) $\hat{\mathbf{c}}$ is in C whenever $\mathbf{c}$ is in C, that is,

$$c_0c_1c_2 \dots c_{n-1} \in C \quad \Longrightarrow \quad c_{n-1}c_0c_1 \dots c_{n-2} \in C.$$

The definition implies that if C is cyclic and $\mathbf{c} \in C$ then the words

$$c_ic_{i+1} \cdots c_{n-1}c_0 \cdots c_{i-1}$$

obtained from $\mathbf{c}$ by any number of cyclic shifts are also in C.

Example 9.6

Which of the following codes are cyclic?

$$C_1 = \{000, 100, 010, 001\}, \qquad C_2 = \{0000, 1010, 0101, 1111\}.$$

Solution A cyclic code must be *linear*, and it must be closed under the cyclic shift operation. Thus C_1 is not a cyclic code, because it is not a linear code ($100 + 010 = 110$, which is not C_1, for example).

On the other hand, C_2 is a cyclic code. It is easy to check both conditions: for example, the cyclic shift of the codeword 1010 is the codeword 0101.

There is a simple way of representing a cyclic shift in algebraic terms. An expression

$$a(x) = a_0 + a_1x + a_2x^2 + \cdots + a_dx^d$$

in which $a_0, a_1, a_2, \ldots, a_d$ are in $\mathbb{F}_2$, is said to be a *polynomial* with *coefficients* in $\mathbb{F}_2$. If $a_d \neq 0$ then the *degree* of $a(x)$ is d. Polynomials can be added and multiplied according to the rules we learn in elementary algebra, and with these rules they form a *ring*, denoted by $\mathbb{F}_2[x]$.

Clearly, there is a correspondence between

the polynomial $a_0 + a_1x + \cdots + a_{n-1}x^{n-1}$ in $\mathbb{F}_2[x]$, and

the word $a_0a_1 \ldots a_{n-1}$ in $\mathbb{F}_2{}^n$.

In this correspondence, the cyclic shift $\hat{\mathbf{a}}$ of $\mathbf{a}$ is represented by the polynomial $\hat{a}(x)$, where

$$\begin{aligned}\hat{a}(x) &= a_{n-1} + a_0x + \cdots + a_{n-2}x^{n-1}\\ &= x(a_0 + a_1x + \cdots + a_{n-1}x^{n-1}) - a_{n-1}(x^n - 1)\\ &= xa(x) - a_{n-1}(x^n - 1).\end{aligned}$$

We shall retain the minus signs for the sake of clarity but, since the coefficients belong to $\mathbb{F}_2$ we could rewrite all the minus signs as plus signs. The result of the calculation is that $\hat{a}(x)$ is equal to $xa(x)$, except for a multiple of $x^n - 1$; in other words

$\hat{a}(x)$ is equal to $xa(x)$ modulo $x^n - 1$.

Suppose we define a new rule for multiplying polynomials, as follows. Given $a(x)$ and $b(x)$, multiply them in the usual way, and then reduce modulo $x^n - 1$. This means that we take $x^n - 1$ to be the same as 0, and so the reduction is equivalent to replacing x^n by 1, x^{n+1} by x, x^{n+2} by x^2, and so on. With this rule it is clear that any polynomial $a(x)$ in $\mathbb{F}_2[x]$ can be reduced to a polynomial with degree less than n. We shall denote by $V^n[x]$ the ring of polynomials with coefficients in $\mathbb{F}_2$, using multiplication modulo $x^n - 1$. Technically, $V^n[x]$ is a *quotient ring* of $\mathbb{F}_2[x]$.

We have constructed the ring $V^n[x]$ so that it is in bijective correspondence with $\mathbb{F}_2{}^n$, specifically $a(x) \leftrightarrow \mathbf{a}$. Furthermore, if $a(x)$ and $b(x)$ correspond to $\mathbf{a}$ and $\mathbf{b}$, then $a(x) + b(x)$ corresponds to $\mathbf{a} + \mathbf{b}$ and $xa(x)$ corresponds to $\hat{\mathbf{a}}$, the first cyclic shift of $\mathbf{a}$.

Example 9.7

Write down the polynomials in $V^6[x]$ that correspond to the words 110101 and 010110, and find their product as elements of $V^6[x]$.

Solution 110101 is represented by $1+x+x^3+x^5$ and 010110 is represented by $x+x^3+x^4$. Multiplying and putting $x^6=1$, $x^7=x$, and so on, we obtain

$$\begin{aligned}
&(1+x+x^3+x^5)(x+x^3+x^4)\\
&=(x+x^3+x^4)+(x^2+x^4+x^5)+(x^4+x^6+x^7)+(x^6+x^8+x^9)\\
&=x+x^2+x^3+x^4+x^5+x^7+x^8+x^9\\
&=x+x^2+x^3+x^4+x^5+1+x+x^2\\
&=1+x^3+x^4+x^5.
\end{aligned}$$

We shall now prove that a cyclic code in ${\mathbb{F}_2}^n$ corresponds to a particular kind of subset of $V^n[x]$. Let R be a ring with commutative multiplication. A subset S of R is said to be an *ideal* if

(i) $a, b \in S \quad \Rightarrow \quad a+b \in S$
(ii) $r \in R$ and $a \in S \quad \Rightarrow \quad ra \in S$.

In other words, an ideal S is closed under addition, and under multiplication *by any element of* R.

Theorem 9.8

A binary code with codewords of length n is cyclic if and only if it corresponds to an ideal in $V^n[x]$.

Proof

Suppose C is a cyclic code, represented as a subset of $V^n[x]$. Since C is linear, if $a(x)$ and $b(x)$ are in C so is $a(x)+b(x)$, and the first condition for an ideal is satisfied. Since $xa(x)$ represents the first cyclic shift of $a(x)$, it follows that $xa(x)$ is in C whenever $a(x)$ is in C. By repeating the same argument $x^i a(x)$ is in C whenever $a(x)$ is in C, for any $i \geq 0$. Any polynomial $p(x)$ is the sum of a number of powers x^i and so, since C is linear, $p(x)a(x)$ is in C. Hence C is an ideal.

Conversely, if C is an ideal, condition (i) tells us that it represents a linear code. Condition (ii) tells us (in particular) that $xa(x)$ is in C whenever $a(x)$ is, so C is cyclic. □

It follows from the theorem that the construction of cyclic codes of length n is equivalent to the construction of ideals in $V^n[x]$. There is a very simple way of constructing such ideals.

Let $f(x)$ be any polynomial in $V^n[x]$. The set of all multiples of $f(x)$ in $V^n[x]$ is clearly an ideal, for if $a(x)$ and $b(x)$ are multiples of $f(x)$, so are

$a(x)+b(x)$ and $p(x)a(x)$ for any $p(x)$. We denote this ideal by $\langle f(x)\rangle$ and refer to it as the ideal *generated* by $f(x)$.

Example 9.9

Construct the ideal generated by $f(x) = 1 + x^2$ in $V^3[x]$, and write down the corresponding code $C \subseteq \mathbb{F}_2{}^3$.

Solution Multiplying $f(x)$ in turn by each element $p(x)$ of $V^3[x]$ and reducing modulo $x^3 - 1$, we obtain the following table.

$p(x)$	$p(x)f(x) \bmod (x^3 - 1)$
0	0
1	$1 + x^2$
x	$1 + x$
$1 + x$	$x + x^2$
x^2	$x + x^2$
$1 + x^2$	$1 + x$
$x + x^2$	$1 + x^2$
$1 + x + x^2$	0

So the ideal $\langle 1 + x^2\rangle$ has just four elements

$$0, \quad 1 + x, \quad x + x^2, \quad 1 + x^2.$$

The corresponding code in $\mathbb{F}_2{}^3$ is $C = \{000,\ 110,\ 011,\ 101\}$.

EXERCISES

9.4. Which of the following codes are cyclic?

$$C_1 = \{0000, 1100, 0110, 0011, 1001\},$$

$$C_2 = \{0000, 1100, 0110, 0011, 1001, 1010, 0101, 1111\}.$$

9.5. Write down the codewords of the cyclic code corresponding to the ideal $\langle 1 + x + x^2\rangle$ in $V^3[x]$, and find a check matrix for this code.

9.6. Show that the ideal $\langle 1 + x\rangle$ in $V^5[x]$ corresponds to the code in $\mathbb{F}_2{}^5$ containing all the words of even weight.

9.7. Does the result in the previous exercise hold when 5 is replaced by an arbitrary integer $n \geq 2$?

9.3 Classification and properties of cyclic codes

The next theorem shows that every cyclic code corresponds to an ideal generated by a polynomial.

Theorem 9.10

Let $C \neq \{0\}$ be a cyclic code (ideal) in $V^n[x]$. Then there is a polynomial $g(x)$ in C such that $C = \langle g(x) \rangle$.

Proof

Since C is not $\{0\}$ it contains a non-zero polynomial $g(x)$ of least degree. If $f(x)$ is any element of C let $q(x)$ and $r(x)$ be the quotient and remainder when $f(x)$ is divided by $g(x)$:

$$f(x) = g(x)q(x) + r(x),$$

where $\deg r(x) < \deg g(x)$. Because $g(x)$ is in the ideal C it follows that $g(x)q(x)$ is in C, and since $f(x)$ is also in C,

$$g(x)q(x) - f(x) = r(x)$$

is in C. This contradicts the definition of $g(x)$ as a non-zero polynomial of least degree in C, unless $r(x) = 0$. Hence $f(x) = g(x)q(x)$, which means that $C = \langle g(x) \rangle$, as claimed. □

The polynomial $g(x)$ obtained in the proof is uniquely determined by the property that it has the least degree in C. For if $g_1(x)$ and $g_2(x)$ both have this property, they have the same degree and their leading coefficients are both 1. (Since the coefficients are in $\mathbb{F}_2$ the only possible coefficients are 0 and 1.) Furthermore, since C is an ideal $g_1(x) - g_2(x)$ is also in C, and its degree is less than the degree of $g_1(x)$ and $g_2(x)$. This is a contradiction unless $g_1(x) - g_2(x) = 0$.

In Example 9.9 we listed the ideal in $V^3[x]$ generated by $1 + x^2$. Inspection of the list shows that in this case the unique non-zero polynomial of least degree is $1 + x$, and Theorem 9.10 tells us that this polynomial is also a generator for the ideal. In general, a cyclic code C will have many generators, but only one of them will have the least degree in C. We shall refer to the unique polynomial with this property as the *canonical generator* of C.

Theorem 9.11

The canonical generator $g(x)$ of a cyclic code C in $V^n[x]$ is a divisor of $x^n - 1$ in $\mathbb{F}_2[x]$.

Proof

Dividing $x^n - 1$ by $g(x)$ in $\mathbb{F}_2[x]$, we obtain a quotient and remainder:

$$x^n - 1 = g(x)h(x) + s(x),$$

where $\deg s(x) < \deg g(x)$. This equation implies that, *in the quotient ring* $V^n[x]$, $s(x) = g(x)h(x)$. Since C is the ideal in $V^n[x]$ generated by $g(x)$ it follows that $s(x)$ is in C. This contradicts the fact that $g(x)$ has the least degree in C, unless $s(x) = 0$. Hence $x^n - 1 = g(x)h(x)$ in $\mathbb{F}_2[x]$, as claimed. □

We shall now explain how the canonical generator $g(x)$ determines the dimension of a cyclic code, and a check matrix for it. Since $g(x)$ is a divisor of $x^n - 1$, we have $g(x)h(x) = x^n - 1$ in $\mathbb{F}_2[x]$. Let

$$g(x) = g_0 + g_1x + \cdots + g_{n-k}x^{n-k}, \qquad h(x) = h_0 + h_1x + \cdots + h_kx^k,$$

where the coefficients g_0, h_0, g_{n-k}, h_k must all be 1, since the product of the polynomials is $x^n - 1$. Let $\mathbf{g}$ be the word

$$\mathbf{g} = g_0g_1 \dots g_{n-k}0\ 0 \dots 0,$$

in ${\mathbb{F}_2}^n$, and let $\mathbf{h}^*$ be the word whose first $k+1$ bits are the coefficients of $h(x)$ in reverse order, followed by $n-k-1$ zeros:

$$\mathbf{h}^* = h_kh_{k-1} \cdots h_0 0\ 0 \cdots 0.$$

Let H be the $(n-k) \times n$ matrix whose rows are $\mathbf{h}^*$ and the first $n-k-1$ cyclic shifts of $\mathbf{h}^*$:

$$H = \begin{pmatrix} h_k & h_{k-1} & & & h_0 & 0 & 0 & & 0 \\ 0 & h_k & \cdot & & h_1 & h_0 & 0 & & 0 \\ 0 & 0 & \cdot & \cdot & h_2 & h_1 & h_0 & & 0 \\ & & & \cdot & \cdot & & & \cdot & \\ & & & & \cdot & \cdot & & & \cdot \\ 0 & 0 & & & h_k & h_{k-1} & & & h_0 \end{pmatrix}.$$

Lemma 9.12

Let $\mathbf{g}_{(i)}$ be the row-vector $0\ 0 \cdots 0 g_0 g_1 \cdots g_{n-k} 0\ 0 \cdots 0$ where there are i zeros at the beginning and $k-1-i$ zeros at the end, that is, the cyclic shift of $\mathbf{g}$ corresponding to $x^i g(x)$ in $V^n[x]$. Then $H\mathbf{g}'_{(i)} = \mathbf{0}'$ $(0 \le i \le n-1)$.

Proof

This follows from the fact that $g(x)h(x) = x^n - 1$ (see Exercise 9.12). □

Theorem 9.13

The matrix H is a check matrix for the cyclic code $C = \langle g(x) \rangle$, and the dimension of C is k.

Proof

A word $\mathbf{c}$ in C corresponds to $c(x) = f(x)g(x)$ in $V^n[x]$. If

$$f(x) = f_0 + f_1 x + \cdots + f_{n-1}x^{n-1},$$

the product $f(x)g(x)$ can be written as

$$c(x) = f_0 g(x) + f_1 x g(x) + \cdots + f_{n-1}x^{n-1}g(x),$$

and so it follows that $\mathbf{c}$ is given by

$$\mathbf{c} = f_0\mathbf{g} + f_1\mathbf{g}_{(1)} + \cdots + f_{n-1}\mathbf{g}_{(n-1)}.$$

By Lemma 9.12, $H\mathbf{g}'_{(i)} = \mathbf{0}'$ for $0 \le i \le n-1$, and so $H\mathbf{c}' = \mathbf{0}'$.

Finally, suppose $\mathbf{y} = y_0 y_1 \cdots y_{n-1}$ is in C. Since $h_0 = 1$, the equation corresponding to the first row of $H\mathbf{y}' = \mathbf{0}'$ is

$$y_k = h_k y_0 + h_{k-1}y_1 + \cdots + h_1 y_{k-1}.$$

So if the values of $y_0, y_1, \ldots, y_{k-1}$ are given, the value of y_k is determined. The equation corresponding to the second row determines y_{k+1} in terms of $y_1, y_2, \ldots, y_k$, and so on. Since there are 2^k possible values for $y_0, y_1, \ldots, y_{k-1}$, we have $|C| = 2^k$, as claimed. □

The foregoing results imply that, in order to find all cyclic codes with word-length n, it suffices to find the irreducible factors of $x^n - 1$ in $\mathbb{F}_2[x]$. It is tedious to find these factors 'by hand', but computer algebra systems such as MAPLE can do it very quickly.

Example 9.14

Given that there are three irreducible factors of $x^7 - 1$ in $\mathbb{F}_2[x]$:

$$x^7 - 1 = (1+x)(1+x+x^3)(1+x^2+x^3),$$

what are the possibilities for a cyclic code with word-length 7?

Solution By combining the three irreducible factors in all possible ways we can obtain $2^3 = 8$ divisors of $x^7 - 1$ in $\mathbb{F}_2[x]$. They are the trivial divisors 1 and $x^7 - 1$, together with

$$1+x, \quad 1+x+x^3, \quad 1+x^2+x^3, \quad (1+x)(1+x+x^3),$$
$$(1+x)(1+x^2+x^3), \quad (1+x+x^3)(1+x^2+x^3).$$

Each of the divisors generates a cyclic code, and (by Theorems 9.10 and 9.11) these are the only cyclic codes of length 7. Clearly $\langle 1 \rangle$ is the code in which every word is a codeword, and $\langle x^7 - 1 \rangle = \langle 0 \rangle$ is the code in which the only codeword is $\mathbf{0}$. The other codes are more interesting. For example, let C be the code with canonical generator $g(x) = 1 + x + x^3$, so that

$$h(x) = (1+x)(1+x^2+x^3) = 1+x+x^2+x^4$$

and $\mathbf{h}^* = 1011100$. It follows from Theorem 9.14 that the dimension of C is 4, and a check matrix for C is

$$\begin{pmatrix} 1 & 0 & 1 & 1 & 1 & 0 & 0 \\ 0 & 1 & 0 & 1 & 1 & 1 & 0 \\ 0 & 0 & 1 & 0 & 1 & 1 & 1 \end{pmatrix}.$$

The seven columns of this matrix are distinct and nonzero, so the code C is equivalent to the Hamming code in ${\mathbb{F}_2}^7$.

EXERCISES

9.8. Write down the factors $x^5 - 1$ in $\mathbb{F}_2[x]$, and hence determine all cyclic codes of length 5.

9.9. Complete the classification of cyclic codes of length 7, along the lines of Example 9.14.

9.10. Show that when n is odd, $x - 1$ occurs exactly once as a factor of $x^n - 1$ in $\mathbb{F}_2[x]$. (It is worth remembering that $x - 1$ and $1 + x$ are the same in $\mathbb{F}_2[x]$.)

9.11. The factorization of $x^{15} - 1$ in $\mathbb{F}_2[x]$ is

$$x^{15}-1 = (1+x)(1+x+x^2)(1+x+x^4)(1+x^3+x^4)(1+x+x^2+x^3+x^4).$$

Using this result, find a canonical generator for a cyclic code equivalent to the Hamming code of length 15.

9.12. Write down the equations for the coefficients of $g(x)$ and $h(x)$ that result from the fact that $g(x)h(x) = x^n - 1$, and hence prove Lemma 9.12.

9.4 Codes that can correct more than one error

The following generalization of Theorem 8.10 is the starting point for the construction of families of codes with arbitrarily large minimum distance.

Theorem 9.15

Let H be a check matrix for a binary code C. Then the minimum distance of C is the equal to the minimum number of linearly dependent columns of H.

Proof

Suppose the minimum distance of C is δ. Then there is a word $\mathbf{c} = c_0c_1 \cdots c_{n-1}$ in C with weight δ (Lemma 8.3). Let $\mathbf{h}'_i$ denote column i of H. The equation $H\mathbf{c}' = \mathbf{0}'$ can be written as

$$c_0\mathbf{h}'_0 + c_1\mathbf{h}'_1 + \ \cdots \ c_{n-1}\mathbf{h}'_{n-1} = \mathbf{0}'.$$

In this equation, δ of the coefficients c_i are equal to 1 and the others are 0. Hence the equation says that δ columns of H are linearly dependent.

Conversely, if a set of columns of H is linearly dependent, the word in which the ith bit is 1 when the ith column is in the set is a codeword. □

Stated positively, the theorem says that a check matrix in which every set of r columns is linearly *independent* defines a code with minimum distance $r+1$, at least. The following lemma shows that a particular kind of matrix has this property, and it will be used to justify the construction given in the next section. Note that the result holds in any field.

Lemma 9.16

Suppose F is a field and $a \in F$ is such that $a^0 = 1, a, a^2, a^3, \ldots, a^{n-1}$ are distinct non-zero elements of F. Let A be the $r \times n$ matrix such that the entry in row i and column j is a^{ij}, where the rows are labelled $i = 1, 2, \ldots, r$ and the columns are labelled $j = 0, 1, \ldots, n-1$. Then any r columns of A are linearly independent over F.

Proof

Let A_S be the $r \times r$ submatrix of A comprising the set S of columns labelled

$s_1, s_2, \ldots, s_r$:

$$A_S = \begin{pmatrix} a^{s_1} & a^{s_2} & . & . & . & a^{s_r} \\ a^{2s_1} & a^{2s_2} & . & . & . & a^{2s_r} \\ & & & & & \\ . & . & . & . & . & . \\ a^{rs_1} & a^{rs_2} & . & . & . & a^{rs_r} \end{pmatrix}.$$

We must show that the columns of A_S are linearly independent, which is equivalent to showing that the determinant of A_S is not zero.

Since a^{s_k} is a factor of every term in column s_k, $\det A_S$ has factors a^{s_1}, a^{s_2}, $\ldots$, a^{s_r}. Also, subtracting column s_k from column s_ℓ produces a column in which every term has $a^{s_\ell} - a^{s_k}$ as a factor. The product of these factors yields a polynomial in a of degree $s_1 + 2s_2 + \cdots + rs_r$. On the other hand, the highest power of a that occurs in the expansion of the determinant is the diagonal term $a^{s_1}a^{2s_2} \cdots a^{rs_r}$, so there are no other factors. Hence

$$\det A_S = \pm a^{s_1+s_2+\cdots+s_r} \prod (a^{s_\ell} - a^{s_k}),$$

where the product is taken over all pairs such that $s_\ell > s_k$. Since it is given that $a^{s_\ell} \neq a^{s_k}$, it follows that $\det A_S \neq 0$. □

EXERCISES

9.13. Without using the argument given in the proof of Lemma 9.16, verify that

$$\det \begin{pmatrix} a^2 & a^5 & a^7 \\ a^4 & a^{10} & a^{14} \\ a^6 & a^{15} & a^{21} \end{pmatrix} = a^{14}(a^5 - a^2)(a^7 - a^2)(a^7 - a^5).$$

9.14. Explain why Theorem 8.10 is a special case of Theorem 9.15.

9.15. Suppose we wish to assign ID numbers in the form of binary words of length n to a set of 16 people, so that the words form a 2-error-correcting linear code – that is a linear code with parameters $(n, 4, 5)$. Show that we shall require n to be at least 10.

9.16. Given the conditions stated in the previous exercise, there is in fact no solution with $n = 10$. [You are not asked to prove this.] However, there is a solution with $n = 11$, which we are going to construct. Let A be the linear code with parameters $(6, 3, 3)$ discussed in Example 8.6 and Section 8.3), and let $B = \{000000, 111111\}$ be the code with parameters $(6, 1, 6)$. Let $C \subseteq \mathbb{F}_2{}^{12}$ be the code formed by words of the form

$$[\mathbf{a} \mid \mathbf{a} + \mathbf{b}] \quad \mathbf{a} \in A,\ \mathbf{b} \in B.$$

Show that C is a linear code with parameters $(12, 4, 6)$ and construct a linear code with parameters $(11, 4, 5)$ by making a suitable modification to C.

9.17. Is the code constructed in the previous exercise a cyclic code?

9.5 Definition of a family of BCH codes

Families of codes that can correct any number of errors were discovered by Bose and Ray-Chaudhuri, and (independently) by Hocqhenghem, in 1959-60. These codes are known as *BCH codes.* Here we shall consider only one case: a family of binary cyclic codes such that, for given integers m and t, the parameters (n, k, δ) satisfy

$$n = 2^m - 1, \quad k \geq n - tm, \quad \delta \geq 2t + 1.$$

Since the BCH codes are cyclic codes, we begin with the factorization of $x^n - 1$ into irreducible polynomials, as in Section 9.3. Suppose it is

$$x^n - 1 \;=\; (x-1)f_1(x)f_2(x)\;\ldots\;f_\ell(x) \quad \text{in } \mathbb{F}_2[x].$$

There is only one possible linear factor $x - 1$, and when n is odd it occurs only once (Exercise 9.10), so the polynomials $f_1(x), \ldots, f_\ell(x)$ all have degree greater than 1. However, if we extend the field then $x^n - 1$ can be split completely into linear factors – in other words, it has n roots. (This procedure is analogous to extending the real field to the complex field, in order to obtain n roots for any polynomial of degree n with real coefficients.)

Our definition of BCH codes will involve the construction of a field E containing $\mathbb{F}_2$, and an element $\alpha \in E$ such that α is a root of the equation $x^n = 1$ in E. Any such α has the property that $\alpha, \alpha^2, \alpha^3, \ldots, \alpha^{n-1}$ are roots of $x^n = 1$. We shall require that these elements of E are distinct, and in that case we shall say that α is a *primitive* root. The construction of α and E will be given shortly, but for the time being we shall proceed on the assumption that it can be done. This means that we can write

$$x^n - 1 \;=\; (x-1)(x-\alpha)(x-\alpha^2)\;\ldots\;(x-\alpha^{n-1}) \quad \text{in } E[x].$$

Lemma 9.17

If α is a primitive root, each of the terms $x - \alpha^i$ $(1 \leq i \leq n-1)$ is a factor of exactly one of the polynomials $f_j(x)$ $(j = 1, 2, \ldots, \ell)$ in $E[x]$.

Proof

Since the coefficients of $f_j(x)$ are in $\mathbb{F}_2$, which is contained in E, we may consider $f_j(x)$ as an element of $E[x]$. Unique factorization holds in $E[x]$, and it follows by comparing the two factorizations of $x^n - 1$ that $f_j(x)$ must be a product of terms of the form $x - \alpha^i$. Since the total degree is the same both cases, a given factor $x - \alpha^i$ can occur in only one $f_j(x)$. □

For $i = 1, 2, \ldots, n-1$ let $m_i(x)$ be the unique irreducible factor $f_j(x)$ of $x^n - 1$ in $\mathbb{F}_2[x]$ for which $x - \alpha^i$ is a factor of $f_j(x)$ in $E[x]$. Equivalently, $m_i(x) = f_j(x)$ when $f_j(x)$ has α^i as a root.

Definition 9.18 (BCH code, designed distance)

Given an odd integer n, let the polynomials $m_i(x) \in \mathbb{F}_2[x]$ be as above, and define $g(x)$ to be the least common multiple of

$$m_1(x),\ m_2(x),\ \ldots\ , m_{dd-1}(x).$$

Then we say that the cyclic code in ${\mathbb{F}_2}^n$ with canonical generator $g(x)$ is a *BCH code with designed distance dd*.

The reason for taking the least common multiple is that the polynomials $m_i(x)$ may not all be distinct. For example, it is easy to show (Exercise 9.18) that for any $f(x) \in \mathbb{F}_2[x]$ we have $f(x)^2 = f(x^2)$. It follows that if α^i is a root of $f_j(x)$ then so is α^{2i}. Hence the polynomials $m_i(x)$ and $m_{2i}(x)$ are the same, and in the definition of $g(x)$ we need only consider the polynomials $m_i(x)$ with i odd. When *dd* is an odd number, the definition of $g(x)$ becomes

$$g(x) = \text{lcm}\{m_1(x), m_3(x),\ \ldots\ , m_{dd-2}(x)\}.$$

We must now address the problem of constructing the primitive root α and the field E. Consider the case $n = 7$, when the irreducible factors of $x^7 - 1$ are

$$x^7 - 1 = (1+x)(1+x+x^3)(1+x^2+x^3).$$

Any root of $x^7 - 1$ must be a root of one of the factors. The only root of $1 + x$ is 1, which is clearly not primitive. If α is a root of $1 + x + x^3$, the equation $1 + \alpha + \alpha^3 = 0$ can be written as $\alpha^3 = 1 + \alpha$. Using this equation, any power of α can be reduced to a polynomial in α with degree not greater than 2. Thus

$$\begin{aligned} \alpha^3 &= 1+\alpha \\ \alpha^4 &= \alpha(1+\alpha) = \alpha+\alpha^2 \\ \alpha^5 &= \alpha(\alpha+\alpha^2) = \alpha^2+\alpha^3 = 1+\alpha+\alpha^2 \\ \alpha^6 &= \alpha(1+\alpha+\alpha^2) = 1+\alpha^2 \\ \alpha^7 &= \alpha(1+\alpha^2) = 1. \end{aligned}$$

Using the correspondence

$$c_0 + c_1\alpha + c_2\alpha^2 \quad \longleftrightarrow \quad c_0\, c_1\, c_2$$

we have

$$\begin{array}{llll} \alpha = 010, & \alpha^2 = 001, & \alpha^3 = 110, & \alpha^4 = 011 \\ \alpha^5 = 111, & \alpha^6 = 101, & \alpha^7 = 100. & \end{array}$$

We have represented the 7 powers of α as the 7 distinct nonzero elements of ${\mathbb{F}_2}^3$. This means that we can take E to be ${\mathbb{F}_2}^3$, with the appropriate operation of multiplication: the product of $a_0a_1a_2$ and $b_0b_1b_2$ is obtained by multiplying the polynomials $a_0 + a_1\alpha + a_2\alpha^2$ and $b_0 + b_1\alpha + b_2\alpha^2$, and using the relation $1 + \alpha + \alpha^3 = 0$ to reduce the answer to a polynomial of the same form.

The same method works whenever $n = 2^m - 1$: it is always possible to find one of the irreducible factors of $x^n - 1$ that defines a primitive root α. The reader who has studied *Galois fields* will recognise the construction of $GF(2^m)$.

To summarize: when $n = 2^m - 1$ the method described above can always be used to construct a field E with 2^m elements, and a primitive root $\alpha \in E$. Furthermore, as we shall see in the next section, the same method produces an explicit check matrix for a BCH code with $n = 2^m - 1$ and a given value of dd.

Here is a simple example, in which the check matrix is obtained directly.

Example 9.19

Take $n = 7$ and α to be a root of $1 + x + x^3$. Find the canonical generator for the BCH code of designed distance $dd = 7$, and show that the code has minimum distance $\delta = 7$.

Solution Since $dd = 7$, the canonical generator is the least common multiple of $m_1(x)$, $m_3(x)$ and $m_5(x)$. It is given that α is a root of $1 + x + x^3$, so $m_1(x) = 1 + x + x^3$.

In order to determine $m_3(x)$ and $m_5(x)$ we need to find the factors that have α^3 and α^5 as roots. Using the representation as elements of ${\mathbb{F}_2}^3$ obtained above we have

$$1 + (\alpha^3)^2 + (\alpha^3)^3 = 1 + \alpha^6 + \alpha^2 = 100 + 101 + 001 = 000,$$

$$1 + (\alpha^5)^2 + (\alpha^5)^3 = 1 + \alpha^3 + \alpha = 100 + 110 + 010 = 000.$$

In other words, both α^3 and α^5 are roots of $1+x^2+x^3$, and $m_3(x) = m_5(x) = 1 + x^2 + x^3$. Hence

$$g(x) = \text{lcm}\{m_1(x), m_3(x), m_5(x)\} = (1 + x + x^3)(1 + x^2 + x^3).$$

Following the rules given in Section 9.3 we find $h(x) = 1 + x$, so the dimension is $k = 1$. The first row of the check matrix is $\mathbf{h}^* = 1100000$ and the other rows are the cyclic shifts 0110000, ..., 0000011. The resulting equations imply that $x_1x_2 \ldots x_7$ is a codeword if and only if $x_1 = x_2 = \cdots x_7$, so the code is $\{0000000, 1111111\}$. Clearly, this code has minimum distance 7.

EXERCISES

9.18. Prove that $f(x)^2 = f(x^2)$ for any polynomial $f(x) \in \mathbb{F}_2[x]$. Hence determine all the polynomials $m_i(x)$, $1 \leq i \leq 6$ when $n = 7$ and the primitive root is a root of $1 + x + x^3$.

9.19. Factorize $x^3 - 1$ in $\mathbb{F}_2[x]$, and hence determine the binary BCH code in ${\mathbb{F}_2}^3$ with $dd = 3$. (Refer to Exercise 9.5.)

9.20. Show that a root of $1 + x^2 + x^3$ can be taken as a primitive root of $x^7 - 1$.

9.6 Properties of the BCH codes

Suppose we know a primitive root α of $x^n - 1$. The following lemma enables us to construct a check matrix for the corresponding BCH code without having to calculate the canonical generator explicitly.

Lemma 9.20

Let $n = 2^m - 1$. If $\mathbf{c} = c_0c_1 \ldots c_{n-1}$ is a codeword in a binary BCH code of designed distance $dd = 2t + 1$, constructed using a primitive root α, then $H^\flat\mathbf{c}' = \mathbf{0}'$, where $H^\flat$ is the $2t \times n$ matrix over E given by

$$H^\flat = \begin{pmatrix} 1 & \alpha & \alpha^2 & \ldots & \alpha^{n-1} \\ 1 & \alpha^2 & \alpha^4 & \ldots & \alpha^{2(n-1)} \\ \cdot & \cdot & \cdot & \cdot & \cdot \\ 1 & \alpha^{2t} & \alpha^{4t} & \ldots & \alpha^{2t(n-1)} \end{pmatrix}.$$

Proof

The polynomial $c(x)$ representing the codeword $\mathbf{c}$ is a multiple of $g(x)$ which, by Definition 9.18, is a multiple of $m_1(x), m_2(x), \ldots, m_{2t}(x)$. Since $m_i(\alpha^i) = 0$ it follows that $c(\alpha^i) = 0$, that is

$$c_0 + c_1\alpha^i + c_2\alpha^{2i} + \ \ldots \ + c_{n-1}\alpha^{i(n-1)} \ = \ 0 \quad (i = 1, 2, \ldots, 2t).$$

These equations are equivalent to the matrix equation $H^\flat \mathbf{c}' = \mathbf{0}'$. □

Theorem 9.21

Let C be a binary BCH code with word-length $n = 2^m - 1$ and designed distance $dd = 2t + 1$. Then the dimension k and the minimum distance δ of C satisfy

$$k \geq n - tm, \qquad \delta \geq dd.$$

Proof

Suppose we have identified a primitive root α of $x^n - 1$ and represented the powers α^i $(1 \leq i \leq n)$ as elements of $\mathbb{F}_2{}^m$. If we replace the entries of the matrix $H^\flat$ by the corresponding words in $\mathbb{F}_2{}^m$ (as *column* vectors), we obtain an $2tm \times n$ matrix H over $\mathbb{F}_2$. The equations

$$c_0 + c_1\alpha^i + c_2\alpha^{2i} + \ \ldots \ + c_{n-1}\alpha^{i(n-1)} \ = \ 0 \quad (i = 1, 2, \ldots, 2t),$$

now correspond to $H\mathbf{c}' = \mathbf{0}'$, so H is a check matrix for C.

The $2tm$ rows of H are not all needed, since some of the polynomials $m_1(x), m_2(x), \ldots, m_{2t}(x)$ coincide. Indeed, we know that $m_i(x)$ and $m_{2i}(x)$ are certainly the same. Thus the number s of distinct polynomials is at most t. It follows that H can be be replaced by an $sm \times n$ matrix, and the dimension of the code is $k = n - sm \geq n - tm$.

Finally, Lemma 9.16 shows that any $2t$ columns of $H^\flat$ are linearly independent, so the same is true for H, and the minimum distance of C is such that $\delta \geq 2t + 1 = dd$. □

Example 9.22

The factorization of $x^{15} - 1$ in $\mathbb{F}_2[x]$ is

$$x^{15} - 1 = (1 + x)(1 + x + x^2)(1 + x + x^4)(1 + x^3 + x^4)(1 + x + x^2 + x^3 + x^4).$$

Show that a root α of $1 + x + x^4$ can be chosen as primitive root, and hence write down a check matrix for the BCH code with $n = 15$ and $dd = 5$.

Solution Since $\alpha^4 = 1 + \alpha$, any power of α can be reduced to a polynomial in α with degree at most 3. Using the correspondence

$$c_0 + c_1\alpha + c_2\alpha^2 + c_3\alpha^3 \longleftrightarrow c_0\,c_1\,c_2\,c_3$$

we find that

$$\begin{array}{llll} \alpha = 0100, & \alpha^2 = 0010, & \alpha^3 = 0001, & \alpha^4 = 1100 \\ \alpha^5 = 0110, & \alpha^6 = 0011, & \alpha^7 = 1101, & \alpha^8 = 1010, \\ \alpha^9 = 0101, & \alpha^{10} = 1110, & \alpha^{11} = 0111, & \alpha^{12} = 1111, \\ \alpha^{13} = 1011, & \alpha^{14} = 1001, & \alpha^{15} = 1000. & \end{array}$$

Since these vectors are all distinct, we conclude that α is a primitive root.

In the case $dd = 5$, $g(x)$ is the lcm of $m_1(x), m_2(x), m_3(x)$. Since $m_2(x) = m_1(x)$ a check matrix for the code is the 8×15 matrix over $\mathbb{F}_2$,

$$H = \begin{pmatrix} 1 & \alpha & \alpha^2 & \alpha^3 & \alpha^4 & \alpha^5 & \alpha^6 & \dots & \alpha^{14} \\ 1 & \alpha^3 & \alpha^6 & \alpha^9 & \alpha^{12} & 1 & \alpha^3 & \dots & \alpha^{12} \end{pmatrix},$$

where α^i stands for the corresponding column vector, as listed above.

According to Theorem 9.21, the code constructed above has $\delta \geq dd = 5$, and so it is a 2-error-correcting code. There is a simple error-correction procedure.

Suppose a word $\mathbf{z} = z_0 z_1 \dots z_{14}$ is received. As usual, we begin by calculating the syndrome $\mathbf{s}$, say $\mathbf{s}' = H\mathbf{z}' = [\mathbf{x}\ \mathbf{y}]'$, where $\mathbf{x}$ and $\mathbf{y}$ are 4-vectors.

- If $\mathbf{x} = \mathbf{y} = 0000$, then $\mathbf{z}$ is a codeword, and there are no bit-errors.
- If, for some i, $\mathbf{x} = \alpha^i$ and $\mathbf{y} = \alpha^{3i}$, the syndrome is the ith column of H, and there is an error in the ith bit of $\mathbf{z}$.
- If it is possible to find i and j satisfying the equations

$$\mathbf{x} = \alpha^i + \alpha^j, \quad \mathbf{y} = \alpha^{3i} + \alpha^{3j},$$

then the syndrome is the sum of the ith and jth columns of H, and $\mathbf{z}$ has errors in the ith and jth bits. So we must solve these equations for i and j. If the equations have no solution, it must be assumed more than two errors have been made.

Example 9.23

Suppose two errors are made in transmission and the syndrome of the received word $\mathbf{z}$ is

$$\mathbf{s}' = H\mathbf{z}' = 1100\,0110'.$$

Which bits are in error?

Solution In this case, $\mathbf{x} = 1100$ and $\mathbf{y} = 0110$. If errors have been made in the ith and jth bits, we have to find i and j such that

$$\alpha^i + \alpha^j = 1100 \qquad \alpha^{3i} + \alpha^{3j} = 0110.$$

One way to do this is to make a list of the pairs (α^i, α^j) for which the first equation is satisfied:

$$(1, \alpha), \quad (\alpha, 1), \quad (\alpha^2, \alpha^{10}) \quad (\alpha^3, \alpha^7), \quad \ldots \quad .$$

Now we can check whether the corresponding pairs satisfy the second equation.

$$1 + \alpha^3 = 1000 + 0001 = 1001,$$
$$\alpha^3 + 1 = 0001 + 1000 = 1001,$$
$$\alpha^6 + 1 = 0011 + 1000 = 1011,$$
$$\alpha^9 + \alpha^6 = 0101 + 0011 = 0110.$$

The conclusion is that bits labelled 3 and 7 are in error, so the intended codeword was

$$\mathbf{c} = \mathbf{z} + 000100010000000.$$

EXERCISES

9.21. Compare the information rate and error-correction properties of the BCH code constructed in Example 9.22 with those of the Hamming code with the same word-length.

9.22. Suppose that the BCH code in Example 9.22 is being used, and a word with syndrome 10101001 is received. Assuming that two bit-errors have been made, which bits should be corrected?

9.23. Investigate the binary BCH codes with $n = 15$ and $dd = 7, 9, 11, 13, 15$, taking a root of $1 + x + x^4$ as a primitive root as in Example 9.22.

Further reading for Chapter 9

Hamming's construction of perfect codes was published in 1950 [**9.4**]. Around the same time Marcel Golay discovered two very remarkable perfect codes that can correct more than one error [**9.2**], and it is now known that these are the only perfect codes with that property. The Golay codes have many links with other interesting structures in group theory, design theory, and geometry.

One of the original motivations for studying cyclic codes was that the shift operations can be implemented by a device known as a shift-register. The general results are mainly due to E. Prange [**9.6**] and W.W. Peterson [**9.5**]. Many families of cyclic codes have been constructed, and they have found numerous practical applications, ranging from space exploration to data processing. The BCH codes were first published in 1960 [**9.1**], and there is now an extensive literature about them: a good introduction is given by Pretzel [**9.7**]. Important advances are still being made in this area – see, for example, Guruswami and Sudan [**9.3**].

9.1 R.C. Bose and D.K. Ray-Chaudhuri. On a class of error correcting codes. *Info. and Control* 3 (1960) 68-79, 279-290.

9.2 M.J.E. Golay. Notes on digital coding. *Proc. IEEE.* 37 (1949) 657.

9.3 V. Guruswami and M. Sudan. Improved decoding of Reed-Solomon and algebraic-geometric codes. *IEEE Trans. Info. Theory* 45 (1999) 1757-1767.

9.4 R.W. Hamming. Error detecting and error correcting codes. *Bell System Tech. J.* 29 (1950) 147-160.

9.5 W.W. Peterson and E.J. Weldon. *Error-Correcting Codes.* MIT Press, Cambridge, Mass. (1972).

9.6 E. Prange. The use of information sets in decoding cyclic codes. *IEEE Trans. Info. Theory.* 8 (1962) 55-59.

9.7 O. Pretzel. *Error-Correcting Codes and Finite Fields.* Oxford University Press, Oxford (1992).

10
Coding natural languages

10.1 Natural languages as sources

Thus far we have studied coding for the purposes of economy and reliability. The third main purpose, security, involves (among other things) the secrecy of messages written in a natural language. For that reason, the complex properties of natural languages play an important part in cryptography.

We begin by discussing how a natural language, specifically English, may be considered as a *source* in the sense of coding theory. Our mathematical model of this source will be referred to as `english`. It is based on the alphabet $\mathbb{A}$ with 27 symbols, the letters `A, B, C, ..., Z`, and the space, denoted by ␣. In `english` there is no distinction between upper and lower case letters, and there are no punctuation marks. For example, the text

Rhett: Frankly my dear, I don't give a damn!

is rendered as the following string in the `english` alphabet.

`RHETT␣FRANKLY␣MY␣DEAR␣I␣DONT␣GIVE␣A␣DAMN`

Although we use the simplified alphabet $\mathbb{A}$, we shall try to ensure that `english` has statistical properties that resemble the real English language.

N. L. Biggs, *An Introduction to Information Communication and Cryptography*,
DOI: 10.1007/978-1-84800-273-9_10,

The fundamental fact is the observation that the frequencies of the symbols are found to be fairly constant over a wide range of texts, ranging from Jane Austen to the *Electronic Journal of Analytical Philosophy.* Typical values of these frequencies, expressed as the number of occurrences per 10000 symbols, are shown in Figure 10.1.

␣	A	B	C	D	E	F	G	H
1753	724	75	231	319	1042	205	126	419

I	J	K	L	M	N	O	P	Q
641	9	53	320	145	554	551	148	6

R	S	T	U	V	W	X	Y	Z
557	668	842	195	91	172	9	133	12

Figure 10.1 A frequency table for symbols in `english`

This *frequency table* defines a probability distribution $\mathbf{p}$ on $\mathbb{A}$. But clearly this table does not tell the whole story about `english`. For example, given that one symbol is the letter `Q`, there is a very high probability that the next symbol is `U`. In other words, it cannot be assumed that `english` is a memoryless source.

In cryptography, a pair of consecutive symbols is known as a *digram*, and the relative frequencies of digrams are significant. For example, it is observed that the digram `UP` occurs more frequently that the digram `UE`. Part of a frequency table for digrams is shown in Figure 10.2. In this table the number in row `B` and column `E` is the number of occurrences of the digram `BE` in a typical piece of text with 10,000 symbols.

	␣	A	B	C	D	E	...
␣	–	212	57	86	69	28	...
A	41	–	3	38	22	–	...
B	–	3	–	–	–	35	...
C	–	41	–	–	–	44	...
D	135	22	–	–	–	54	...
E	365	57	–	16	73	24	...
...	...	...	...	...	...	...	...

Figure 10.2 Part of a frequency table for digrams in `english`

In mathematical terms, the frequency table for digrams defines a probability distribution $\mathbf{p}^2$ on $\mathbb{A}^2$. It is clear that for any ordered pair of symbols ij, the probability $\mathbf{p}^2(ij)$ is not equal to $\mathbf{p}(i)\mathbf{p}(j)$. For example, according to the table given above

$$\mathbf{p}(\mathtt{A}) = 0.0724 \text{ and } \mathbf{p}(\mathtt{B}) = 0.0075, \quad \text{so} \quad \mathbf{p}(\mathtt{A})\mathbf{p}(\mathtt{B}) \approx 0.0006,$$

whereas $\mathbf{p}^2(\mathtt{AB})$ is approximately 0.0003.

EXERCISES

10.1. If a frequency table for digrams is given, how can a frequency table for the individual symbols be obtained?

10.2. On the basis of the frequency table shown in Figure 10.1, estimate the average length of a word in `english`.

10.2 The uncertainty of `english`

Recall that a *stationary source* has the property that the probability of finding any given sequence of consecutive symbols is the same at all points of the emitted stream. If we think of a novel as a stream of symbols emitted by the source called `english`, the stationary property means that the probability of finding `BOOT` (for example) on page 1 is the same as finding it on page 99, or on any other page.

Formally, we require that for each $n \geq 1$ there is a probability distribution $\mathbf{p}^n$ on the set of n-tuples $\mathbb{A}^n$. If the stream emitted by the source is represented by the sequence of random variables $\xi_1\xi_2\xi_3 \cdots$, then for any $k \geq 1$ and for any n-tuple of symbols $x_1x_2\ldots x_n \in \mathbb{A}^n$ we require that

$$\Pr(\xi_{k+1} = x_1,\ \xi_{k+2} = x_2,\ \ldots,\ \xi_{k+n} = x_n) = \mathbf{p}^n(x_1x_2\cdots x_n).$$

It is reasonable to assume that the statistical properties of `english`, and any other natural language, can be modelled by representing it as a stationary source. For the time being we shall focus on the implications of this assumption, remembering that it may not cover all the features of `english`. In particular, we recall that there is good definition of the *entropy* of a stationary source (Definition 4.8).

In the case of a natural language we shall be concerned (initially at least) with coding a message in the source alphabet by a string of symbols in the same alphabet. For that reason it is appropriate to measure the entropy in terms of logarithms to base b, where b is the number of symbols in the alphabet. For example, in `english` $b = 27$. In order to emphasize this point, we shall use the word *uncertainty* in this context.

Definition 10.1 (Uncertainty of a natural language)

Suppose we regard a natural language with an alphabet of size b as a stationary source represented by distributions $\mathbf{p}^n$ $(n \geq 1)$. Let $\mathcal{U}_n = H_b(\mathbf{p}^n)/n$. Then the *uncertainty* of the language is defined to be

$$\mathcal{U} = \inf_{n \in \mathbb{N}} \mathcal{U}_n.$$

Our choice of units means that the number

$$\mathcal{U}_n = \frac{H_b(\mathbf{p}^n)}{n} = \frac{H_2(\mathbf{p}^n)}{n \log_2 b}$$

lies between 0 and 1. It can be thought of as the average uncertainty per symbol, when the source is considered as emitting a stream of blocks of size n.

Example 10.2

A language has three symbols a, b, c, and the frequency table for digrams is

	a	b	c
a	0.22	0.08	0.10
b	0.10	0.16	0.04
c	0.08	0.06	0.16

If the language is considered as a stationary source, what are the values of $\mathcal{U}_1$ and $\mathcal{U}_2$?

Solution Using the the addition formula for $\mathbf{p}^1$ in terms of $\mathbf{p}^2$ (Section 4.3) we have

$$\mathbf{p}^1(a) = 0.22 + 0.08 + 0.10 = 0.4,$$
$$\mathbf{p}^1(b) = 0.10 + 0.16 + 0.04 = 0.3,$$
$$\mathbf{p}^1(c) = 0.08 + 0.06 + 0.16 = 0.3.$$

Hence

$$\mathcal{U}_1 = H_3(\mathbf{p}^1) = 0.4 \log_3(1/0.4) + 0.3 \log_3(1/0.3) + 0.3 \log_3(1/0.3) \approx 0.9912.$$

For $\mathcal{U}_2$ we have

$$\mathcal{U}_2 = \frac{1}{2}H_3(\mathbf{p}^2) = 0.5\Big(0.22\log_3(1/0.22) + 0.08\log_3(1/0.08) + \cdots$$

$$+ \cdots + 0.16\log_3(1/0.16)\Big) \approx 0.9474.$$

The fact that $\mathcal{U}_2 < \mathcal{U}_1$ reflects the fact that the uncertainty is less when the relationship between consecutive symbols is taken into account.

On the assumption that `english` is a stationary source, we can attempt to estimate its uncertainty by using experimental data. We start with the *first-order approximation*, consisting of the distribution $\mathbf{p}^1$ on $\mathbb{A}$, as measured by the frequency table (Figure 10.1). This results in the estimate

$$\mathcal{U}_1 = H_{27}(\mathbf{p}^1) = \sum_{i\in\mathbb{A}} \mathbf{p}^1(i)\log_{27}(1/\mathbf{p}^1(i)),$$

which works out at about 0.85. The *second-order approximation* is determined by the frequency table for digrams, which defines a distribution $\mathbf{p}^2$ on the set $\mathbb{A}^2$. This gives the estimate

$$\mathcal{U}_2 = \frac{1}{2}H_{27}(\mathbf{p}^2) = \frac{1}{2}\sum_{ij\in\mathbb{A}^2} \mathbf{p}^2(ij)\log_{27}(1/\mathbf{p}^2(ij)).$$

After a rather long calculation, a value of about 0.70 is obtained.

Unfortunately, computing reliable estimates $\mathcal{U}_n$ for larger values of n is very difficult, because there are 27^n individual contributions to the entropy, and most of them are very small. Nevertheless, it is reasonable to expect that, as longer sequences are taken into account, the estimate of uncertainty per symbol will decrease, because common words such as `AND`, `THE`, and `THIS` will contribute significantly to the results. On the other hand, at this level it is hard to justify the claim that every human author represents the common source that we should like to call `english`. For example, some authors avoid the use of certain four-letter words, while others are less constrained.

For the record, we simply note that the conventional assumption, based on extensive calculations, is that

$$\mathcal{U} = \inf_{n\in\mathbb{N}} \mathcal{U}_n \approx 0.3.$$

Roughly speaking, if a long piece of text is given, then we can predict what comes next with about 70% certainty. In the next section we shall look at this result from a different point of view.

EXERCISES

10.3. The language `footish` is mainly used by professional footballers. The language uses a set of three symbols, $\{* \quad ! \quad ?\}$. An example of a string recently emitted is as follows.

$$?! * *?!! * *??!! * *?! * *?$$

Estimate the uncertainty of the first-order and second-order approximations to `footish`.

10.4. What is the value of $\mathcal{U}_1$ for a language in which all the symbols are equally probable?

10.5. Consider a language with two symbols x, y, such that the frequency table for digrams is as follows.

	x	y
x	0.3	0.2
y	0.2	0.3

Show that $\mathcal{U}_1 = 1$ but $\mathcal{U} < 1$ for this language.

10.3 Redundancy and meaning

There are good reasons for believing that the property of being a stationary source does not capture all the features of a natural language. In addition to its non-trivial statistical properties, a natural language has another fundamental property, which we may loosely refer to as *meaning*. Although it is plain that the function of natural language is to convey meaningful messages, it is not easy to express this idea in mathematical terms.

One difficulty is that it is quite possible to generate strings of symbols that make no sense, although they have statistical properties very similar to the source that we have called `english`. Suppose we take a book, assumed to be a typical output of `english`. If we rearrange the pages of the book, and the sentences on each page, we shall have an output that is virtually indistinguishable (in statistical terms) from the original book. But clearly, it makes no sense.

One characteristic property of a meaningful message is that it is possible to shorten it without destroying the meaning. At a very basic level that is the reason why we can use *abbreviations*, such as LSE and USA. Similarly, omitting some letters from a meaningful message will not prevent it being understood.

For example, 7 of the original 29 symbols have been omitte

ITIIGINGTRAITOMORW

but the meaning has not been lost. This property of a natural la
known as *redundancy*. We now give a heuristic argument, due to Shanno
links the redundancy of a language with its uncertainty.

Consider the set of meaningful messages of length n emitted by the sourc
that we call `english`. Suppose that it has been found by experiment that the proportion of symbols that can be omitted from these messages without destroying the meaning is f_n. The process of removing nf_n symbols can be thought of as encoding the message of length n by a message in the same alphabet, but with length

$$\ell_n = n(1 - f_n).$$

Now the coding theorem for stationary sources (Theorem 4.11) says that the optimum value of ℓ_n/n is close to $\mathcal{U}$, the uncertainty of the source. Thus it reasonable to assume that

$$\inf_{n \in \mathbb{N}} f_n = \inf_{n \in \mathbb{N}} (1 - \ell_n/n) = 1 - \mathcal{U}.$$

Shannon's argument suggests that it is reasonable to make the following definition.

Definition 10.3 (Redundancy)

The *redundancy* of a natural language with b symbols is defined to be

$$\mathcal{R} = 1 - \mathcal{U},$$

where $\mathcal{U}$ is the uncertainty as in Definition 10.1.

There are several ways of looking at this definition. One of them is to regard it as an alternative means of calculating $\mathcal{U}$, using experimental results on redundancy. Most experiments suggest that well over half of the letters in a long message can be omitted, although naturally the results vary according to the rule that is used to suppress the letters. This observation is consistent with the estimate $\mathcal{U} \approx 0.3$ given in the previous section.

In summary, there are several ways of estimating the uncertainty and redundancy of a natural language such as `english`. However, neither linguistic theorists nor mathematicians have succeeded in formulating a theory that captures all the subtleties of natural language, and so numerical estimates are very imprecise. More important is the fact that the concepts of redundancy and meaning are highly relevant to the practical aspects of cryptography, as we shall see in the next section.

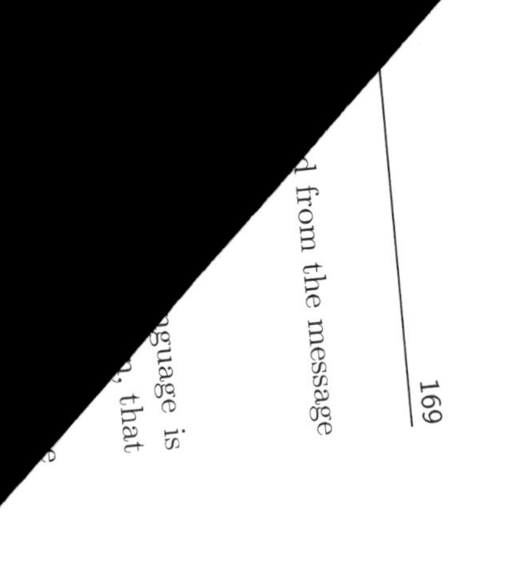

igure 10.1) to estimate approximately the
ssage in **english** that consists of vowels

english messages, in which the vowels
emoved.

) DNTCRYFRMRGNTN

on has been told that the redundancy
In their unceasing quest for efficiency,
ation in a simplistic way by sending the
message to all staff. What was their rule for suppressing
symbols, and what is the intended message?

PES RMME T CMLT YU RPRS BFR TE ED O TR.

10.9. The widespread use of mobile phones for sending text messages has led to a new language, which we may call **textish**. Discuss the relationship between **textish** and **english**. Which language has the greater uncertainty?

10.4 Introduction to cryptography

We are now ready to discuss the traditional aspect of cryptography: the study of coding for the purpose of transmitting secret messages written in a natural language. Traditionally, secrecy was required mainly in diplomatic and military communications, but nowadays it plays an important part in our everyday lives, for example, in managing our financial affairs.

Partly because it has a long history, cryptography has developed its own terminology. We shall begin with a brief description of a few of the methods that have been used in the past, and by doing so we shall be able to introduce most of the special terminology of the subject. We shall also encounter many of the peculiar problems that arise in this area.

One of the oldest cryptographic systems is said to have been used by Julius Caesar over two thousand years ago. It was discussed briefly in Section 1.6 of this book. For a message emitted by the idealized source that we have called **english**, the system is as follows. The sender chooses a number k between 1 and 25 and replaces each letter by the one that is k places later, in alphabetical order, with the obvious rule for letters at the end of the alphabet. In order to simplify the explanation we shall assume that the space ⊔ is not changed. (In practice the aim is obviously *not* to make things simple, and a different rule would be used.) For example, when $k = 5$ the letters are replaced as follows.

```
⊔ A B C D E F G H I J K L M N O P Q R S T U V W X Y Z
⊔ F G H I J K L M N O P Q R S T U V W X Y Z A B C D E
```

Using this rule, the message

SEE⊔YOU⊔TOMORROW   becomes   XJJ⊔DTZ⊔YTRTWWTB   .

In cryptography this process is known as *encryption*, and the number k is the *key* ($k = 5$ in the example above). In mathematical terms, encryption is simply coding, with the additional feature that the coding function E_k depends upon a parameter k. For example, in our version of Caesar's system E_k is the function $\mathbb{A} \to \mathbb{A}$ defined by

$$E_k(x) = [x + k] \quad (x \neq \sqcup), \qquad E_k(\sqcup) = \sqcup,$$

where $[x + k]$ stands for the symbol k places after x, with the obvious rule for the letters at the end of the alphabet. The function E_k is extended so that it acts on messages (strings of symbols) by concatenation: for example

$$E_k(\texttt{CAT}) = E_k(\texttt{C})E_k(\texttt{A})E_k(\texttt{T}).$$

As usual, we require that the extended E_k is an injection, so that the coding is uniquely decodable. (Definition 1.13). The general situation is as follows.

Definition 10.4 (Encryption functions)

Let $\mathcal{M}$ and $\mathcal{C}$ be sets of messages, and let $\mathcal{K}$ be a set. Suppose that for each $k \in \mathcal{K}$ there is an injective function $E_k : \mathcal{M} \to \mathcal{C}$. Then we say that $\{E_k\}$ is a set of *encryption functions*, and an element $k \in \mathcal{K}$ is called a *key*.

In Caesar's system $\mathcal{M}$ and $\mathcal{C}$ are both taken to be the set $\mathbb{A}^*$ of strings of symbols in $\mathbb{A}$, and $\mathcal{K}$ is the set of numbers in the range $1 \leq k \leq 25$.

The main lesson from the history of cryptography is that it is futile to try to conceal the method of encryption, that is, the form of the functions E_k. If a number of people wish to communicate securely, they cannot hope to conceal the system that they are using. The security of the system must depend only on the specific value they agree to assign to the key k.

Nowadays it is customary to give human names, Alice, Bob, Eve, to the participants in security procedures. (In practice, the participants are machines of various kinds.) Roughly speaking, Alice is the sender of a message, Bob is the receiver, and Eve is an adversary who is trying to intervene in some way. Within this framework we can distinguish several requirements that come under the general heading of security.

The most obvious requirement is *Confidentiality*: a message from Alice to Bob should not be understood by Eve. This is the requirement that we shall consider in this chapter. However, modern cryptography is also concerned with other aspects of security, such as *Authenticity*, *Integrity*, and *Non-repudiation*. We shall discuss these matters in Chapter 14.

Let us assume that Alice and Bob have agreed on a method of encryption and the value of the key k, so that they both know the encryption function E_k. The message that Alice wishes to send Bob is known as *plaintext*, and for the time being we shall assume that is expressed in a natural language, such as `english`. Alice uses the function E_k to transform the plaintext into coded form, known as *ciphertext*, and sends this to Bob.

So if Alice and Bob have agreed to use the Caesar system with $k = 5$, and the plaintext is

`SEE␣YOU␣TOMORROW`

then Bob will receive the ciphertext

`XJJ␣DTZ␣YTRTWWTB` .

In this case it is fairly obvious how Bob should retrieve the plaintext, but it is helpful to consider the situation in more general terms.

Suppose we are given a set of encryption functions $E_k : \mathcal{M} \to \mathcal{C}$. According to Definition 10.4 each E_k is an injective function. This means that if $E_k(m) = c$, then m is the *unique* plaintext that E_k encrypts as c. Consequently there is a function F, defined on the image of E_k, that takes c to m:

$$F(E_k(m)) = m \qquad \text{for all } m \in \mathcal{M}.$$

Formally, F is the *left-inverse* of E_k.

In practical cryptography it is not enough to know that a left-inverse exists, it is necessary to have an explicit rule for calculating it. That is the point of the following definition.

Definition 10.5 (Decryption functions)

Let $\{E_k\}$ $(k \in \mathcal{K})$ be a set of encryption functions $\mathcal{M} \to \mathcal{C}$. Then the set of functions $\{D_\ell\}$ $(\ell \in \mathcal{K})$ is a set of *decryption functions* for $\{E_k\}$ if for each $k \in \mathcal{K}$ there is an $\ell \in \mathcal{K}$ such that D_ℓ is the left-inverse of E_k:

$$D_\ell(E_k(m)) = m \qquad \text{for all } m \in \mathcal{M}.$$

In the Caesar system, we take ℓ such that $\ell = -k \pmod{26}$ and the left-inverse of E_k is

$$D_\ell(x) = [x + \ell] \quad (x \neq \sqcup), \qquad D_\ell(\sqcup) = \sqcup.$$

Since Bob is assumed to know k, he also knows ℓ, and he can decrypt the message from Alice. Note that in this case the decryption function D_ℓ and the encryption function E_k have the same form, although that is not necessary in general.

Consider now the situation when Eve intercepts the ciphertext

XJJ⊔DTZ⊔YTRTWWTB .

She cannot simply apply a decryption function and recover the plaintext, because she does not know which key has been used. If she wishes to obtain the plaintext, the obvious method is to try to find the key.

The process of obtaining the plaintext corresponding to some ciphertext, either by finding the value of the key k or by some indirect method, is said to be *breaking* the system. Any method which (Eve hopes) will achieve this is known as an *attack*. We have already observed that Eve may be assumed to know which system is being used; consequently, if she finds the key, then she knows how to use it.

For the Caesar system there is a simple attack by the method known as *exhaustive search*. Because the only possible values of k are $1, 2, 3, \ldots, 25$, it is easy to try each of them in turn. Assuming that the plaintext is a message expressed in a natural language, Eve will know when the right key is found, because the message will be 'meaningful'.

Example 10.6

Suppose Eve intercepts the ciphertext

SGZN⊔OY⊔MUUJ⊔LUX⊔EUA .

What is the corresponding plaintext?

Solution When the keys $k = 1, 2, 3, \ldots$ are tried in turn on the initial part of the ciphertext, the result is meaningless until $k = 6$ is reached.

$k = 1$	RFYM⊔	$\cdots$
$k = 2$	QEXL⊔	$\cdots$
$k = 3$	PDWK⊔	$\cdots$
$k = 4$	OCVJ⊔	$\cdots$
$k = 5$	NBUI⊔	$\cdots$
$k = 6$	MATH⊔	$\cdots$

So, it is worth trying the key $k = 6$ on the rest of the message. This produces a meaningful message, and the plaintext is

MATH␣IS␣GOOD␣FOR␣YOU .

EXERCISES

10.10. Find the key that has been used to produce the following ciphertext. (Here a space is represented by a blank.)

VJKU KU CP GZEGNNGPV EQWTUG CPF VJG NGEVWTGT
UJQWNF TGEGKXG C DKI RCA TKUG

10.11. The following ciphertext has been produced by a variant of the Caesar system in which the letters are divided into blocks of three for convenience, and the spaces between the words of the plaintext have been ignored. Find the key and the message.

VJC QNV JCR LBR BXO CNW DBN ODU

10.12. The ciphertext CTGPC is the result of encrypting a meaningful word using the Caesar system. Show that there are two possible words satisfying this description.

10.5 Frequency analysis

How can Alice and Bob defend against the attack by exhaustive search? One obvious idea is to increase the number of keys. Caesar's system uses the very simple rule $x \mapsto [x + k]$ for replacing the letters, and there are only 25 possible values of the key k. But clearly *any* permutation of the 26 letters can be used. There are 26! such permutations, and

$$26! \approx 4 \times 10^{27},$$

so Eve will require substantial resources if she wishes to try all of them in turn.

In cryptography a system that simply permutes the letters of the alphabet in a specific way is known as *mono-alphabetic substitution.* The key is a permutation σ of the 26 letters, and the encryption function can be defined as

$$E_\sigma(x) = \sigma(x) \quad (x \neq \sqcup), \qquad E_\sigma(\sqcup) = \sqcup.$$

(In practice, the space would probably be permuted as well, but we shall use this definition in order to simplify the exposition.) Clearly, the left-inverse of

E_σ is the function D_τ that uses the inverse permutation $\tau = \sigma^{-1}$:

$$D_\tau(x) = \tau(x) \quad (x \neq \sqcup), \qquad D_\tau(\sqcup) = \sqcup.$$

It is often convenient to use a key that can be memorized, and this can be done by choosing a *keyword*. For example, if the keyword `PERSONALITY` is chosen, the corresponding permutation σ is defined as follows.

```
⊔ A B C D E F G H I J K L M N O P Q R S T U V W X Y Z
⊔ P E R S O N A L I T Y B C D F G H J K M Q U V W X Z
```

The permutation can be written compactly in *cycle notation*:

$$\sigma = (\sqcup)(\mathtt{APG})(\mathtt{BEOFNDSKYXWVUQHL})(\mathtt{CRJTM})(\mathtt{I})(\mathtt{Z}).$$

One advantage of this notation is that is easy to write down the inverse permutation:

$$\sigma^{-1} = (\sqcup)(\mathtt{AGP})(\mathtt{BLHQUVWYXYKSDNFOE})(\mathtt{CMTJR})(\mathtt{I})(\mathtt{Z}).$$

Although the attack by exhaustive search requires large resources, another method of attack, known as *frequency analysis*, is usually more effective. The method was used by Arab cryptographers in the ninth century AD. It is based on the fact that a plaintext message in a natural language will have the statistical properties discussed in Section 10.1. Furthermore, it is reasonable to assume that the text conveys a meaningful message.

It follows that, in any message of reasonable length, the symbols, digrams, and so on, will occur with frequencies close to those given in the standard frequency tables. Thus if the symbol x occurs with frequency n_x in the plaintext, and the encryption function is E_σ, the symbol $\sigma(x)$ will occur with frequency n_x in the ciphertext. Using the data provided by this observation, and the fact that the plaintext is meaningful, the ciphertext can be decrypted.

In summary, the attack by frequency analysis on a piece of ciphertext produced by mono-alphabetic substitution is carried out as follows:

- count the frequencies of the symbols in the ciphertext;
- compare these with the standard frequencies given in the tables;
- test the likely correspondences of the symbols, until a meaningful plaintext is obtained.

The third step involves making and testing hypotheses, and there is no definite rule for how that should be done. The following example is based on a very crude application of the method. Obviously, more sophisticated arguments would be used in reality.

Example 10.7

Suppose Eve intercepts some ciphertext, part of which is

. . . XPYVBH VWX WROCZBVPYH . . . .

(The space symbol ⊔ has been omitted because we are assuming, for the sake of exposition, that the spaces are unaltered.) Eve believes that Alice has sent this to Bob, using a mono-alphabetic substitution E_σ. She has counted the number of occurrences of each letter in a sample of 1000 symbols from the ciphertext, and the most frequent letters are

Y	Z	V	R	H	W	P	C	X	. . .
97	79	64	63	60	54	47	40	37	. . .

What is the plaintext?

Solution Comparing the results with the frequency table in Figure 10.1, Eve would start by assuming that the letters Y and Z represent E and T, the two most common letters in **english**, in that order.

Similarly, the next most frequent letters, V, R, H, can be assumed to represent A, I, and S, in some order. To decide which order, Eve must test alternative hypotheses on a segment of the ciphertext, such as that displayed above. She has assumed Y $\mapsto$ E and Z $\mapsto$ T, and she might use the frequency table for digrams to decide that YV is more likely to represent EA than EI. Also, noting that a word is more likely to end in S than I, it would be reasonable to try

V $\mapsto$ A R $\mapsto$ I H $\mapsto$ S.

In that case the selected part of the ciphertext would be decrypted as follows:

XPYVBH VWX WROCZBVPYH
..EA.S A.. .I..T.A.ES .

Eve will now look at the next most frequent letter, W, and assume that it stands for O or N. After a few trials she will find that the most likely correspondence is W $\mapsto$ N. This gives

XPYVBH VWX WROCZBVPYH
..EA.S AN. NI..T.A.ES .

Now it is easy to deduce the remaining correspondences, such as X $\mapsto$ D, . . . , and complete the decryption. Note that, in order to find the plaintext, Eve does not have to obtain the full key (the permutation σ of $\mathbb{A}$), only enough of it to give a meaningful message.

EXERCISES

10.13. Write down in cycle notation the permutation of $\mathbb{A}$ determined by the keyword `REPUBLICAN`. Use this key to encrypt the message

`HAPPY BIRTHDAY GRANNY` .

10.14. Suppose you have obtained some ciphertext, encrypted by a mono-alphabetic substitution, with the spaces unaltered. The number of occurrences of the symbols that occur most frequently in a sample of about 1000 letters in the ciphertext is:

A	B	C	D	E	F	I	K	P	S	X	Z
55	58	40	73	44	51	54	3	62	96	15	30

Part of the ciphertext is

`DA XS AF BAD DA XS DEPD CI DES KZSIDCAB` .

Find the corresponding plaintext.

10.15 Alice and Bob are exchanging messages with a mono-alphabetic substitution system. For simplicity, they wish to use *exactly* the same key for both encryption and decryption, while ensuring that only the space is unaltered. Write down a suitable key in cycle notation. If Eve knows that they are using a key with this property, is it possible for her to carry out an attack by exhaustive search?

Further reading for Chapter 10

'Natural language' is a very broad concept, and it must be stressed that our discussion refers only to language in *written* form. Linguists and mathematicians have not yet succeeded in formulating rules that distinguish meaningful messages from nonsense, and for that reason we must tread carefully when we assign numerical values to concepts such as redundancy. The most successful approach so far is the experimental one suggested by Shannon [**10.2**], an excellent account of which is given by Welsh [**10.4**, Chapter 6].

The original reference to Caesar's system is by Suetonius in the second century AD [**10.3**]. The ninth century Arab manuscript on frequency analysis is described in an article by Al-Kadi [**10.1**]. Other classical systems are described in the book by Singh listed at the end of Chapter 1 [**1.3**].

10.1 I.A. Al-Kadi. The origins of cryptology: the Arab contribution. *Cryptologia* 16 (1992) 97-126.

10.2 C. Shannon. Prediction and entropy of printed English. *Bell Syst. Tech. J.* 30 (1951) 50-64.

10.3 Suetonius. *Lives of the Caesars, LVI.*

10.4 D. Welsh. *Codes and Cryptography.* Oxford University Press (1988).

The development of crypt

11.1 Symmetric key cryptosystems

In the previous chapter we described a framework for cryptography based on a set of plaintext messages $\mathcal{M}$, a set of ciphertext messages $\mathcal{C}$, and a set of keys $\mathcal{K}$. For each $k \in \mathcal{K}$ there is an encryption function $E_k : \mathcal{M} \to \mathcal{C}$ with a left-inverse, the decryption function D_ℓ. In other words,

$$D_\ell(E_k(m)) = m \qquad \text{for all } m \in \mathcal{M}.$$

We shall refer to this framework as a *cryptosystem*.

Cryptosystems can be constructed and implemented in many ways. In the mono-alphabetic substitution systems discussed in Chapter 10, the corresponding keys k and ℓ are inverse permutations, and so they are related in a very simple way. Indeed, for many centuries it was tacitly assumed that any practical cryptosystem must have a similar property. It is now clear that the assumption is unjustified, but systems with this property are still widely used, and we give them a name.

Definition 11.1 (Symmetric key cryptosystem)

Suppose we have a cryptosystem in which D_ℓ is the left-inverse of E_k. The system is said to be a *symmetric key cryptosystem* if, whenever either one of k, ℓ is known, then it is easy to calculate the other one.

N. L. Biggs, *An Introduction to Information Communication and Cryptography*,
DOI: 10.1007/978-1-84800-273-9_11,

The word 'easy' can be interpreted as indicating that the calculation can be successfully carried out on any reasonably efficient computer. We shall say more about this point in due course.

EXERCISES

11.1 Suppose we are using a mono-alphabetic substitution system, as in Section 10.5, based on the keyword `DEMOCRAT`. Write down the key for encryption as a permutation in cycle notation, and derive the key for decryption in the same form.

11.2 Poly-alphabetic encryption

For many hundreds of years cryptographers were mainly concerned with ways of making their systems safe against the attack by frequency analysis. In a mono-alphabetic system, with a given key, the plaintext letter `E` (for example) is always replaced by the same letter in the ciphertext, say `X`. Consequently `X` will be distinctive in the ciphertext, because it will almost certainly be the most frequently occurring letter. This weakness can be avoided by using a system in which `E` is not always replaced by the same letter.

Definition 11.2 (Poly-alphabetic encryption)

In a *poly-alphabetic* system the rule for encryption is that each plaintext letter is replaced by a letter that depends, not only on the letter itself, but also its position in the text.

One simple method of poly-alphabetic encryption was known as long ago as the 16th century, and is usually known as the *Vigenère system.* It uses only the 'Caesar' permutations of the letters – that is the cyclic shifts of step-size k $(1 \leq k \leq 25)$, denoted in the previous chapter by $E_k(x) = [x+k]$. As usual, for the purposes of exposition we ignore the spaces, in this case by deleting them from the plaintext altogether.

In the Vigenère system the key is a sequence $K = (k_1, k_2, \ldots, k_m)$ of numbers. If the plaintext is $x_1x_2x_3 \ldots x_i \ldots$, the rule for encryption is that

$$E_K(x_i) = [x_i + k_j] \quad \text{if } i \equiv j \pmod{m}.$$

For example, if the key has length $m = 4$, then $x_1, x_5, x_9, \ldots$ are encrypted with a shift of k_1, $x_2, x_6, x_{10}, \ldots$ are encrypted with a shift of k_2, and so on. This is obviously a symmetric key system, since decryption can be done by reversing the shifts.

In practice the key K is usually expressed as a keyword, the letters of the keyword representing the relevant shifts.

Example 11.3

Suppose the keyword is `CHANGE`, representing the key

$$K = (3, 8, 1, 14, 7, 5).$$

What is the encrypted form of the following plaintext?

`THEPROOFOFTHEPUDDINGISINTHEEATING`

Solution The letters in positions $1, 7, 13, 19, 25, 31, 37$ are `T,O,E,N,T,I`, and they are encrypted with a shift of 3, becoming `W,R,H,Q,W,L`. The other letters are encrypted in a similar way. The entire process can be set out as follows.

T	H	E	P	R	O	O	F	O	F	T	H	E	P	U	D	D	I	N	G	...
3	8	1	14	7	5	3	8	1	14	7	5	3	8	1	14	7	5	3	8	...
W	P	F	D	Y	T	R	N	P	T	A	M	H	X	V	R	K	N	Q	O	...

It is clear that the Vigenère system has the desired effect of smoothing out the frequencies of the letters in the ciphertext. For example, suppose that a piece of `english` plaintext contains letters with their usual frequencies (Figure 10.1) and the keyword is `CHANGE`, as in the example above. Then the ciphertext letter `A` represents one of the letters `X S Z M T V`, and hence its expected frequency is

$$\frac{1}{6}(9 + 668 + 12 + 145 + 842 + 91) \approx 294.$$

Similarly, the letter `Z` in the ciphertext represents one of the letters `W R Y L S U`, so its expected frequency is

$$\frac{1}{6}(172 + 557 + 133 + 320 + 668 + 195) \approx 341.$$

These numbers are much closer than would be expected if a mono-alphabetic substitution were used. (Note that since we are ignoring spaces, which occur approximately 1753 times in every 10000 symbols, the frequency of each letter is here expressed as a proportion of 8247 symbols, rather than 10000. But this does not affect the conclusion.)

For several centuries the Vigenère system was thought to be unbreakable, and was referred to as *le chiffre indéciffrable* – the undecipherable cipher. However, the system has an inherent weakness, because the choice of the shifts must be based on a definite rule, otherwise the intended receiver will not be able to decrypt. This weakness can be exploited to mount a successful attack.

The basic strategy is simple. If the *length* m of the keyword is known, then the letters occurring in positions $1, m+1, 2m+1, \ldots$ are all encrypted with the same shift k_1, and hence k_1 can be found by standard frequency analysis. The same holds for the letters in positions $i, m+i, 2m+i, \ldots$, for any i. Hence if m is known, finding the keyword itself presents no difficulty, provided that a reasonably large chunk of ciphertext is available.

It was not until the 19th century that an effective method of finding m was discovered. The method has since been refined, and it is now almost automatic. However, a significant amount of calculation is required and a realistic example would require more space than is available here. But the basic idea is easy to understand. Take a piece of ciphertext c and translate it forward by ℓ places, so that the ith symbol in c is the $(i+\ell)$th symbol in the new text. If ℓ is not equal to m (or a multiple of m) there will be no correlation between the two texts. But if ℓ is equal to m (or a multiple of m) then corresponding symbols have been encrypted by the same substitution, and some correlation can be expected. It turns out that a simple statistic, the *index of coincidence*, measures the degree of correlation. Thus, provided that a reasonable sample of ciphertext is available, the value of m will be revealed by calculating this statistic for a range of values of ℓ.

The important general conclusion is that no cryptosystem can be guaranteed secure against unforeseen methods of attack.

EXERCISES

11.2. Complete the encryption of the message in Example 11.3.

11.3. The ciphertext

`MKWJAFQCHIUPVPONWHFDRKEFIROFEHGQMRGM`

is the result of encrypting a message in the Vigènere system, using the keyword `SCRAMBLE`. Find the message.

11.4. Suppose Vigènere encryption with the keyword `CHANGE` is applied to a piece of plaintext in which the frequencies of the letters are as in **`english`** (Figure 10.1). What is the frequency of the letters `D` and `Q` in the ciphertext? What would be the frequency of each letter (per 8247 letters) if the distribution were uniform?

11.3 The Playfair system

Although it was thought to be unbreakable, the Vigenère system was difficult to use. Consequently there was considerable interest in alternative methods. One such method was the *Playfair system*, in which the digrams are permuted, rather than the individual letters.

A key for the system is derived from a 5×5 square containing 25 letters. (In order to make a square, J is identified with I, and spaces are ignored.) Double-letter digrams of the form LL are not allowed, so the number of digrams available is $25 \times 24 = 600$. The number of keys is therefore 600!, which is enormous, and finding the key by exhaustive search is unlikely to be feasible.

The arrangement of the letters in the square may be random, or it may be defined by a keyword. Figure 11.1 depicts the square based on the keyword PERSONALITY. Provided the sender and receiver can both remember the keyword, and the simple rules given below, the encryption and decryption procedures are straightforward.

```
P E R S O
N A L I T
Y B C D F
G H K M Q
U V W X Z
```

Figure 11.1 The Playfair square with the keyword PERSONALITY

In the encryption process spaces are ignored, and doubled letters are separated by inserting a dummy letter, such as X or Z. The plaintext is then split into digrams, and each digram xy is encrypted according to the following rules. (Minor variations are possible.)

Case 1 Suppose x and y are in different rows and different columns. Then $xy \mapsto ab$ where x, y, a, b are the corners of a rectangle, and x, a are in the same row.

Case 2 Suppose x, y are in the same row. Then $xy \mapsto uv$, where u and v are the letters immediately to the right of x and y respectively. (If one of x and y is at the end of the row, the letter at the beginning of the row is used.)

Case 3 Suppose x, y are in the same column. Then $xy \mapsto uv$, where u and v are the letters immediately below x and y respectively. (If one of x and y is at the bottom of the column, the letter at the top of the column is used.)
For example,

$$\mathtt{MA} \mapsto \mathtt{HI} \quad \mathtt{BW} \mapsto \mathtt{CV} \quad \mathtt{YF} \mapsto \mathtt{BY} \quad \mathtt{SD} \mapsto \mathtt{IM}\,.$$

The Playfair system is a symmetric key system. Both the encryption key and the decryption key are permutations of the set of 600 digrams; Alice uses a Playfair square and the rules given above to construct the key for encryption, and Bob uses the same square, but slightly different rules (Exercise 11.7) to construct the key for decryption. The two keys are inverse permutations.

Example 11.4

Suppose you receive the following ciphertext, knowing that the Playfair system has been used and the keyword is PERSONALITY. What is the plaintext?

AQLAGCPZQTOLVAMLIYHSISEIHP

Solution The ciphertext broken up into a sequence of digrams is

AQ LA GC PZ QT OL VA ML IY HS IS EI HP .

Using the rules in reverse, and the Playfair square shown in Figure 11.1, the corresponding sequence of digrams is

TH AN KY OU FO RT HE KI ND ME SX SA GE .

Remembering the conventions about spaces and doubled-letters, it is clear that the plaintext is

THANK YOU FOR THE KIND MESSAGE .

As might be expected, it is possible to attack and break the Playfair system by frequency analysis. The attack uses the frequency table for digrams, and follows the same lines as the method using the frequency table for single symbols (Section 10.5). However, this procedure obviously requires more data than single-symbol analysis, and it takes longer to complete. Thus the system can provide an acceptable level of security in certain circumstances.

EXERCISES

11.5. Construct the Playfair square with keyword WHYSCRAMBLING, and use it to encrypt the plaintext

THERE WILL BE A MEETING OF THE GROUP TOMORROW .

11.6. Decrypt the following message, which has been encrypted by the Playfair rules with the same keyword as in the previous exercise.

GMANBYZIIFCXWDHQNKENHLDUGKWRFMFUPGMH

11.7. Write down explicitly the rules for decryption corresponding to the rules for encryption given in the text above.

11.4 Mathematical algorithms in cryptography

Up to this point we have used mathematics mainly for the purpose of explaining how certain cryptosystems worked. In the twentieth century it gradually became clear that mathematics could be employed in a more fundamental way.

We have already used the basic idea when we faced the problem of constructing codes with good error-correction properties. Thus, in Chapter 8 we allowed the symbols 0 and 1 in the binary alphabet $\mathbb{B}$ to have algebraic properties. We introduced the notation $\mathbb{F}_2$ for the *field* whose elements are 0 and 1 and ${\mathbb{F}_2}^n$ for the *vector space* of n-tuples over $\mathbb{F}_2$. The use of these algebraic constructions greatly extended the range of methods that could be employed.

In general, the symbols in a message can be represented by the elements of an algebraic structure in many ways. The simplest way to represent a set of n objects by an algebraic structure is to use the *integers mod* n, denoted by $\mathbb{Z}_n$. For example, the 27-symbol alphabet $\mathbb{A}$ could be represented by $0, 1, 2, \ldots, 26$, regarded as elements of $\mathbb{Z}_{27}$. We shall assume that the reader is familiar with fact that the integers mod n can be added and multiplied in such a way that they satisfy the usual rules of arithmetic. Technically, we say that $\mathbb{Z}_n$ is a *ring*.

In fact, it is often convenient to use an alphabet with a prime number of symbols, because when p is a prime number the integers mod p form a *field*. This means that every non-zero element has a multiplicative inverse. To emphasize its special nature we shall denote this field by $\mathbb{F}_p$. Thus if we extend the alphabet $\mathbb{A}$ by allowing messages to contain commas and full-stops, in addition to the usual 27 symbols, then there are 29 symbols, a prime number. We can represent the space ␣ by 0, the letters `A`, `B`, ..., `Z` by $1, 2, \ldots, 26$, and the comma and full-stop by 27 and 28, all considered as elements of the field $\mathbb{F}_{29}$. For example, if the message is

`I␣CAME,␣I␣SAW,␣I␣CONQUERED.`

then we replace it by

$$9\ 0\ 3\ 1\ 13\ 5\ 27\ 0\ 9\ 0\ 13\ 1\ 23\ 27\ 0\ 9\ 0\ 3\ 15\ 14\ 17\ 21\ 5\ 18\ 5\ 4\ 28 \quad .$$

The point is that, although arithmetic had no meaning when the symbols were 'just' symbols, we can now perform all the operations of elementary arithmetic, including division by a non-zero quantity. Furthermore, there are good algorithms for performing these operations, based on the methods we learned in elementary school. This possibility vastly increases the range of encryption methods that are available.

A further possibility arises when we recall the standard technique of splitting the stream of symbols into blocks of an appropriate size m, so that each block becomes an m-vector over $\mathbb{F}_{29}$. Now, not only can we 'do arithmetic' on the individual symbols, we can use linear algebra to manipulate the vectors.

Example 11.5

Represent the message `I␣CAME,␣I␣SAW,␣I␣CONQUERED.` as a string of 2-vectors over $\mathbb{F}_{29}$, and find the string of 2-vectors that results when the matrix

$$K = \begin{pmatrix} 3 & 4 \\ 2 & 3 \end{pmatrix}$$

is applied. How could the resulting string be transformed back into the original one?

Solution The message is represented by

$$[9\ 0]'\ [3\ 1]'\ [13\ 5]'\ \cdots \quad .$$

The dashes indicate that we shall regard the blocks as column vectors. Since

$$\begin{pmatrix} 3 & 4 \\ 2 & 3 \end{pmatrix} \begin{pmatrix} 9 \\ 0 \end{pmatrix} = \begin{pmatrix} 27 \\ 18 \end{pmatrix},$$

and so on, this string is transformed into

$$[27\ 18]'\ [13\ 9]'\ [1\ 12]'\ \cdots \quad .$$

Given the second string, we can recover the first one by applying the inverse matrix, which in this case is

$$K^{-1} = (3 \times 3 - 4 \times 2)^{-1} \begin{pmatrix} 3 & -4 \\ -2 & 3 \end{pmatrix} = \begin{pmatrix} 3 & 25 \\ 27 & 3 \end{pmatrix}.$$

The example illustrates the basic idea behind an early attempt to use abstract mathematical techniques in cryptography. In two papers published around 1930, L.S. Hill suggested that messages in the form given above could be encrypted by applying a linear transformation – that is, by multiplying the vectors by a matrix. Generally, in Hill's system the key is an invertible $m \times m$ matrix K and the encryption function is

$$E_K(\mathbf{x}) = K\mathbf{x},$$

where $\mathbf{x}$ is an m-vector. If $\mathbf{y} = K\mathbf{x}$, then $K^{-1}\mathbf{y} = \mathbf{x}$. Hence the decryption function corresponding to E_K is given by $D_L(\mathbf{y}) = L\mathbf{y}$, where $L = K^{-1}$. Hill's sytem is another example of a symmetric key system, since K^{-1} can be computed from K using the standard methods of matrix algebra.

EXERCISES

11.8. Represent the message

```
AN␣APPLE␣A␣DAY␣KEEPS␣DOCTOR␣AWAY.
```

as a string of elements of ${\mathbb{F}_{29}}^2$.

11.9. Continuing from the previous exercise, calculate the first four blocks of the string that results when the plaintext is encrypted using the key

$$K = \begin{pmatrix} 5 & 3 \\ 3 & 2 \end{pmatrix}.$$

11.5 Methods of attack

How might Hill's system be attacked? An attack by exhaustive search would require checking all the $m \times m$ invertible matrices over $\mathbb{F}_{29}$. An $m \times m$ matrix has m^2 components, and if each component is an element of $\mathbb{F}_{29}$, the number of possibilities is 29^{m^2}. Almost all these matrices are invertible, and consequently m can be chosen so that an attack by exhaustive search is not feasible.

If exhaustive search is not feasible, the standard approach to a symmetric key cryptosystem is a *known ciphertext* (or *ciphertext-only*) attack. Eve obtains a piece of ciphertext c and tries to use this information in order to find the keys for encryption k and decryption ℓ. If these keys are found, Eve can not only decrypt the known ciphertext, using the rule $D_\ell(c) = m$, but also any other ciphertexts that are sent using the same keys.

The attack by frequency analysis is a known ciphertext attack. But in the Hill system this attack is prevented by the fact that the letters are thoroughly scrambled. Any given sequence of letters will be encrypted in many different ways, depending on its position within a block and the other letters that occur in that block (see Exercise 11.11).

However, a significant weakness of Hill's system is the possibility of another form of attack: it may be possible for Eve to obtain some pieces of plaintext m and the corresponding ciphertexts c. Cryptographers recognise two kinds of attack based on this information.

- In a *known plaintext* attack, Eve has obtained some pairs (m_i, c_i), where each m_i is plaintext and c_i is the corresponding ciphertext.
- In a *chosen plaintext* attack, Eve has obtained the ciphertexts c_i corresponding to a number of specific plaintexts m_i that she has chosen.

In practice, the details will depend on the circumstances.

Example 11.6

Suppose Eve intercepts some ciphertext which (she believes) contains a report of an ambush, using the Hill system with $m = 2$. She suspects that the ciphertext

$$[15\ 15]'\ [25\ 15]'\ [17\ 3]'$$

represents `AMBUSH`. How can she test her hypothesis and find the key?

Solution In 2-vectors over $\mathbb{F}_{29}$ the word `AMBUSH` is represented by

$$[1\ 13]'\ [2\ 21]'\ [19\ 8]'.$$

Denote the key by $K = \begin{pmatrix} a & b \\ c & d \end{pmatrix}$. Eve's hypothesis is that the sequence $[1\ 13]'\ [2\ 21]'\ [19\ 8]'$ is encrypted as $[15\ 15]'\ [25\ 15]'\ [17\ 3]'$. If this is true, the blocks `AM` and `BU` are encrypted as

$$\begin{pmatrix} 15 \\ 15 \end{pmatrix} = \begin{pmatrix} a & b \\ c & d \end{pmatrix}\begin{pmatrix} 1 \\ 13 \end{pmatrix}, \quad \begin{pmatrix} 25 \\ 15 \end{pmatrix} = \begin{pmatrix} a & b \\ c & d \end{pmatrix}\begin{pmatrix} 2 \\ 21 \end{pmatrix}.$$

Consequently

$$\begin{pmatrix} 15 & 25 \\ 15 & 15 \end{pmatrix} = \begin{pmatrix} a & b \\ c & d \end{pmatrix}\begin{pmatrix} 1 & 2 \\ 13 & 21 \end{pmatrix},$$

$$\begin{pmatrix} a & b \\ c & d \end{pmatrix} = \begin{pmatrix} 15 & 25 \\ 15 & 15 \end{pmatrix}\begin{pmatrix} 1 & 2 \\ 13 & 21 \end{pmatrix}^{-1}.$$

Provided Eve can do some elementary algebra in $\mathbb{F}_{29}$, she can calculate that

$$K = \begin{pmatrix} a & b \\ c & d \end{pmatrix} = \begin{pmatrix} 2 & 1 \\ 5 & 3 \end{pmatrix}.$$

Eve can test her hypothesis first by checking that, with this K, the third block `SH` is encrypted as she suspects:

$$\begin{pmatrix} 2 & 1 \\ 5 & 3 \end{pmatrix}\begin{pmatrix} 19 \\ 8 \end{pmatrix} = \begin{pmatrix} 17 \\ 3 \end{pmatrix}.$$

Since this works, she can then apply

$$K^{-1} = \begin{pmatrix} 3 & -1 \\ -5 & 2 \end{pmatrix}$$

to the rest of the ciphertext. If a meaningful plaintext results, the problem is solved. If not, Eve must try a different piece of ciphertext.

EXERCISES

11.10. Eve routinely intercepts messages encrypted using the Hill system with blocks of size 2, and a key that is changed each day. The first message every day is a weather forecast that begins

FORECAST␣FOR ... .

If today's message begins with the ciphertext [13 7]′ [20 9]′, what is today's key? [Hint: in order to invert a matrix with determinant $\Delta \in \mathbb{F}_{29}$ you must find the inverse of Δ in $\mathbb{F}_{29}$.]

11.11. Suppose the Hill system is being used with blocks of size 5, and the key is the following matrix over $\mathbb{F}_{29}$:

$$K = \begin{pmatrix} 0 & 1 & 1 & 1 & 1 \\ 1 & 0 & 1 & 1 & 1 \\ 1 & 1 & 0 & 1 & 1 \\ 1 & 1 & 1 & 0 & 1 \\ 1 & 1 & 1 & 1 & 0 \end{pmatrix}.$$

Show that the blocks AMPLE, DAMES, CRAMP, DREAM are encrypted in such a way that a different pair of elements of $\mathbb{F}_{29}$ appears in the positions corresponding to AM in each case.

11.12. Suppose Alice and Bob communicate using the Hill system with $m = 3$. What information must Eve obtain in order to make a successful known plaintext attack? If Eve wishes to use a chosen plaintext attack, suggest a good choice for the plaintexts.

11.13. Consider a cryptosystem in which the symbols are represented by elements of a field $\mathbb{F}_p$ and each symbol is encrypted by applying an encryption function of the form

$$E_{\alpha,\beta}(x) = \alpha x + \beta \quad (\alpha, \beta \in \mathbb{F}_p,\ \alpha \neq 0).$$

The key is the pair (α, β). Write down a suitable decryption function and explain why this is a symmetric key system. Explain how the system could be broken by a known plaintext attack. If it were possible to use chosen plaintext, what choice(s) would be be simplest?

Further reading for Chapter 11

At this point the reader is advised to refer again to the historical accounts of cryptography listed at the end of Chapter 1. These books contain a great deal of valuable information on traditional cryptosystems, such as the poly-alphabetic ones, and the methods that have been used to attack them.

The method of coincidences that is now used to break a Vigènere cipher is largely the work of William Friedmann [**11.1**]. Hill's system was described by him in two papers in 1929 and 1931 [**11.2**], [**11.3**].

11.1 W.F. Friedman. *The Index of Coincidence and its Applications in Cryptology.* Dept. of Ciphers Publ. 22, Illinois, 1922.

11.2 L.S. Hill. Cryptography in an algebraic alphabet. *Amer. Math. Monthly* 36 (1929) 306-311.

11.3 L.S. Hill. Concerning certain linear transformation apparatus of cryptography. *Amer. Math. Monthly* 38 (1931) 135-154.

12
Cryptography in theory and practice

12.1 Encryption in terms of a channel

The continued failure to construct an unbreakable cryptosystem led to attempts to analyse the process of encryption in mathematical terms. Shannon proposed a framework based upon the idea that converting plaintext into ciphertext can be regarded as transmission through a channel.

Let us assume that there is a finite set $\mathcal{M}$ of plaintext messages that might be sent. Denote by p_m the probability that the message is m: in other words, there is a probability distribution $\mathbf{p}$ on $\mathcal{M}$. Note the implicit assumption that p_m is not zero: messages that will never be sent are not included in $\mathcal{M}$. We shall regard $(\mathcal{M}, \mathbf{p})$ as a memoryless source, which is the input to a 'channel', as follows.

Let the set of possible keys be $\mathcal{K}$, and let E_k be the encryption function for the key $k \in \mathcal{K}$. Suppose that the probability that E_k is used is r_k, so that we have a probability distribution $\mathbf{r}$ on the set $\mathcal{K}$. It is assumed that the choice of k is independent of the message, m. The output of the system is the ciphertext $c = E_k(m)$, hence the probability that c occurs is

$$q_c \;=\; \Pr(\text{ciphertext is } c) \;=\; \sum p_m r_k,$$

where the sum is over the set of pairs (m, k) such that $E_k(m) = c$. So we have a probability distribution $\mathbf{q}$ on the set $\mathcal{C}$, and we can think of $(\mathcal{C}, \mathbf{q})$ as the output of a channel (Figure 12.1).

N. L. Biggs, *An Introduction to Information Communication and Cryptography*,
DOI: 10.1007/978-1-84800-273-9_12,

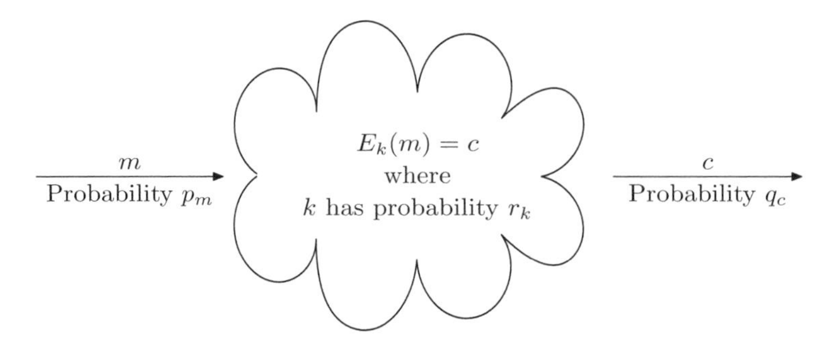

Figure 12.1 Encryption as transmission through a channel

If we regard $\mathbf{p}$ and $\mathbf{q}$ as row vectors in the usual way (see Chapter 5), the transformation effected by the encryption process is defined by a channel matrix Γ such that $\mathbf{q} = \mathbf{p}\Gamma$. Comparison with the formula for q_c given above shows that we can define Γ as follows:

$$\Gamma_{mc} = \Pr(c \mid m) = \sum_k r_k,$$

where the sum is taken over all keys k such that $E_k(m) = c$. Note that Γ depends on the distribution $\mathbf{r}$, just as (for example) the channel matrix for the extended BSC depends upon the bit-error probability e.

In reality, the sets $\mathcal{M}$, $\mathcal{C}$, and $\mathcal{K}$ will be very large, but here is a small example for the purposes of illustration.

Example 12.1

Let $\mathcal{M} = \mathcal{C}$ be the vector space ${\mathbb{F}_2}^2$ of ordered pairs xy, and take $\mathcal{K} = \{1, 2\}$ with $r_1 = r$, $r_2 = 1 - r$. Suppose the encryption functions are

$$E_1(xy) = xy + 01, \qquad E_2(xy) = xy + 10.$$

If the input distribution $\mathbf{p}$ on ${\mathbb{F}_2}^2$ is given by

00	01	10	11
0.1	0.2	0.3	0.4

what is the output distribution $\mathbf{q}$?

Solution Noting that in this example there is at most one k for each pair (m, c), and arranging the rows and columns in the order $00, 01, 10, 11$, the channel matrix Γ is

$$\begin{pmatrix} 0 & r & 1-r & 0 \\ r & 0 & 0 & 1-r \\ 1-r & 0 & 0 & r \\ 0 & 1-r & r & 0 \end{pmatrix}.$$

Hence the output distribution $\mathbf{q} = \mathbf{p}\Gamma$ is

$$\begin{array}{cccc} 00 & 01 & 10 & 11 \\ 0.3 - 0.1r & 0.4 - 0.3r & 0.1 + 0.3r & 0.2 + 0.1r \end{array}.$$

The basic method of attacking a cryptosystem is the *known ciphertext attack* (Section 11.5), in which Eve obtains a piece of ciphertext c and attempts to find the key k. Thus we are led to consider the uncertainty about k given c, or in the terms used in Chapter 5, the *conditional entropy* $H(\mathbf{r} \mid \mathbf{q})$. In this context it has a special name.

Definition 12.2 (Key equivocation)

Given probability distributions $\mathbf{p}$ on $\mathcal{M}$ and $\mathbf{r}$ on $\mathcal{K}$, the *key equivocation* of the system represented by Γ is $H(\mathbf{r} \mid \mathbf{q})$, where $\mathbf{q} = \mathbf{p}\Gamma$.

The following lemma expresses the key equivocation in slightly simpler terms.

Lemma 12.3

In the notation used above, the key equivocation is given by

$$H(\mathbf{r} \mid \mathbf{q}) \;=\; H(\mathbf{r}) + H(\mathbf{p}) - H(\mathbf{q}).$$

Proof

By definition, $H(\mathbf{r} \mid \mathbf{q}) = H(\mathbf{j}) - H(\mathbf{q})$, where $\mathbf{j}$ is the distribution of the pair (k, c). The entropy of $\mathbf{j}$ is the same as the entropy of the distribution of (k, m), since k and m determine c. Since the distributions of k and m are independent, this is equal to $H(\mathbf{r}) + H(\mathbf{p})$. Hence

$$H(\mathbf{r} \mid \mathbf{q}) = H(\mathbf{j}) - H(\mathbf{q}) = H(\mathbf{r}) + H(\mathbf{p}) - H(\mathbf{q}).$$

□

As a first application of the channel formalism we consider the case where $\mathcal{M}$ is a set of meaningful plaintext messages in a natural language. Although our assumptions may appear rather optimistic, the results do provide a useful insight.

We consider a system in which plaintext and ciphertext messages are strings in an alphabet X, and each plaintext is encrypted as a ciphertext of the same length, say n. In other words, $\mathcal{M} = \mathcal{C} = X^n$. Provided that n is not too small ($n = 10$ will usually suffice), the following assumptions about the distributions $\mathbf{p}$ and $\mathbf{q}$ on X^n are reasonable.

1. The entropy per symbol of the distribution on plaintexts, $H(\mathbf{p})/n$, is approximately $\mathcal{U} \log_2 |X|$, where $\mathcal{U}$ is the uncertainty of the language (Definition 10.1).

2. All ciphertexts of length n are equally probable, so the entropy $H(\mathbf{q})$ is $\log_2 |X^n| = n \log_2 |X|$.

In this situation the key equivocation is given approximately by

$$H(\mathbf{r} \mid \mathbf{q}) \approx H(\mathbf{r}) + n(\mathcal{U} - 1) \log_2 |X| = H(\mathbf{r}) - n\mathcal{R} \log_2 |X|,$$

where $\mathcal{R}$ is the redundancy of the language (Definition 10.3).

As n increases, the right-hand side decreases, and there is a value n_0 at which it becomes zero. The value n_0 is called the *unicity point*. It can be thought of as the length of a piece of ciphertext c that is sufficient to ensure there is no uncertainty: that is, there is only one message-key pair that could produce c. Using the estimate given above, we have

$$n_0 \approx \frac{H(\mathbf{r})}{\mathcal{R} \log_2 |X|}.$$

It is possible to obtain numerical estimates for n_0 by inserting some (rather speculative) values for the quantities on the right-hand side, but the significant conclusion is simply that a unicity point exists.

EXERCISES

12.1. Evaluate the key equivocation in Example 12.1 when $r = 0.5$.

12.2. Repeat the previous exercise with $r = 0.51$. Is it true that the key equivocation is greatest when $r = 0.5$?

12.3. Given the ciphertext c, let $K(c)$ denote the set of keys k for which there is a plaintext m such that $E_k(m) = c$. If $|K(c)| > 1$ then all but one of the keys in $K(c)$ is said to be *spurious*. In the Caesar system spurious keys can exist for ciphertexts of length 5 – see Exercise 10.12. Construct an example of a ciphertext of length 3 for which a spurious key exists.

12.4. Show that, in the Caesar system, there is a spurious key for a ciphertext of length 4, one of the plaintexts being the place-name ADEN.

12.2 Perfect secrecy

It is reasonable to say that *perfect secrecy* occurs when the uncertainty about a piece of plaintext is not altered if the corresponding ciphertext is known. In the channel formalism, this translates into the fact that the uncertainty associated with the input source $(\mathcal{M}, \mathbf{p})$ should be the same as its uncertainty when the output source $(\mathcal{C}, \mathbf{q})$ is known. The latter quantity is the conditional entropy of $\mathbf{p}$ with respect to $\mathbf{q}$, and as in Section 5.3 we denote it by

$$H(\Gamma; \mathbf{p}) = H(\mathbf{p} \mid \mathbf{q}) \quad \text{when } \mathbf{q} = \mathbf{p}\Gamma.$$

Definition 12.4 (Perfect secrecy)

Suppose that a cryptosystem is defined by a set of keys $\mathcal{K}$ and encryption functions $E_k : \mathcal{M} \to \mathcal{C}$ for $k \in \mathcal{K}$. Suppose also that a probability distribution $\mathbf{r}$ on $\mathcal{K}$ is given. Then the system is said to have *perfect secrecy* if

$$H(\mathbf{p}) = H(\Gamma; \mathbf{p}) \quad \text{for all distributions } \mathbf{p} \text{ on } \mathcal{M},$$

where Γ is the channel matrix for the distribution $\mathbf{r}$.

Theorem 12.5

A cryptosystem has perfect secrecy if and only if, for all probability distributions $\mathbf{p}$ on the set of plaintexts $\mathcal{M}$, the channel matrix Γ for the distribution $\mathbf{r}$ on the set of keys $\mathcal{K}$ satisfies

$$\Gamma_{mc} = q_c \quad \text{for all } m \in \mathcal{M} \text{ and } c \in \mathcal{C},$$

where $\mathbf{q}$ is the corresponding distribution on the ciphertext space $\mathcal{C}$.

Proof

According to Theorem 5.11, $H(\mathbf{p}) = H(\Gamma; \mathbf{p})$ if and only if $\mathbf{p}$ and $\mathbf{q} = \mathbf{p}\Gamma$ are independent. This means that the probability t_{mc} that the plaintext m and the ciphertext c occur is equal to $p_m q_c$. But we also have the equation

$$t_{mc} = \Pr(c \mid m)\, p_m = \Gamma_{mc} p_m,$$

so $t_{mc} = p_m q_c$ if and only if $\Gamma_{mc} = q_c$, given our assumption that no messages with zero probability are included in $\mathcal{M}$. □

The theorem suggests that perfect secrecy is very rare. The entries Γ_{mc} of the channel matrix are fixed, whereas the probabilities q_c vary with the input distribution $\mathbf{p}$. Hence the equality $\Gamma_{mc} = q_c$ can only hold in very special circumstances. An example will be given in the next section, but first here is a simple condition that follows directly from the theorem.

Corollary 12.6

If a cryptosystem has perfect secrecy then the number of keys is at least equal to the number of messages: $|\mathcal{K}| \geq |\mathcal{M}|$.

Proof

Fix $c \in \mathcal{C}$ such that $q_c > 0$. Then the condition $\Gamma_{mc} = q_c$ implies that, for each m, the sum of r_k taken over all keys k such that $E_k(m) = c$ is not zero. In particular, for each m there is such a k. Since E_k is an injection, it follows that the keys for different m's must be different. Hence there are at least as many keys as messages. □

EXERCISES

12.5. Show that the system described in Example 12.1 does not have perfect secrecy, for any value of r.

12.6. Let $\mathcal{M} = \mathcal{C}$ be the vector space ${\mathbb{F}_2}^2$, and suppose the key space $\mathcal{K}$ is also ${\mathbb{F}_2}^2$, with each key being equally probable. Suppose the encryption function for the key $\alpha\beta \in {\mathbb{F}_2}^2$ is

$$E_{\alpha\beta}(xy) = xy + \alpha\beta.$$

Show that this system has perfect secrecy. [This is a special case of the general result proved in the next section.]

12.3 The one-time pad

Perfect secrecy is a very restrictive condition, and examples are rare. But there is one important example, known as the *one-time pad.* In this system the plaintext and ciphertext sets $\mathcal{M}$ and $\mathcal{C}$ are both sets of n-tuples in a given alphabet, such as the `english` alphabet $\mathbb{A}$. Since the theory works for any alphabet, for the purposes of exposition we shall use the binary alphabet $\mathbb{F}_2$.

We take $\mathcal{M} = \mathcal{C} = {\mathbb{F}_2}^n$, the vector space of strings of length n over $\mathbb{F}_2$. The set of keys $\mathcal{K}$ is also ${\mathbb{F}_2}^n$, so Corollary 12.6 is satisfied. For each $k \in \mathcal{K}$ the encryption function E_k is

$$m \mapsto m + k \quad (m \in \mathcal{M}).$$

The corresponding decryption function is the same function, $c \mapsto c+k$, because

$$(m + k) + k = m.$$

Clearly this is a symmetric key system.

The idea behind the system is that the plaintext is expressed as a string m of n bits, and is encrypted by adding to it an arbitrary key string k of the same length. The next theorem says that the system has perfect secrecy if the key string is chosen uniformly at random. In other words, each new message is encrypted by using a new key, all keys being equally probable. That is the reason for the name *one-time pad.*

Theorem 12.7

The cryptosystem described above, with the probability distribution on the keys defined by

$$r_k = \frac{1}{2^n} \quad (k \in \mathcal{K}),$$

has perfect secrecy.

Proof

For any $m \in \mathcal{M}$ and $c \in \mathcal{C}$ there is a unique k such that $E_k(m) = c$, specifically $k = m + c$. It follows that the entries of the 'channel matrix' Γ are given by

$$\Gamma_{mc} = r_{m+c} = \frac{1}{2^n} \quad \text{for all } m, c.$$

Therefore the output distribution $\mathbf{q}$ is defined in terms of the input distribution $\mathbf{p}$ by

$$q_c = \sum_m p_m \Gamma_{mc} = \frac{1}{2^n} \sum_m p_m = \frac{1}{2^n}.$$

Since $q_c = \Gamma_{mc}$ for all m and c, and for all $\mathbf{p}$, it follows from Theorem 11.7 that the system has perfect secrecy. □

EXERCISES

12.7. Describe briefly how the one-time pad system could be applied to messages in the 27-symbol `english` alphabet $\mathbb{A}$. Illustrate your answer by decrypting the message RQOQXMMBDHK, given that the key is MATHEMATICS.

12.8. Suppose that a user of the one-time pad makes a mistake, and sends two different messages m_1 and m_2 using the same key. If Eve intercepts the ciphertexts c_1 and c_2, what information does this provide?

12.9. Continuing from the previous exercise, suppose Eve suspects that somewhere in m_1 a certain bit-string occurs – it might be the string representing TUESDAY for example. How can Eve test this hypothesis?

12.4 Iterative methods

Throughout the twentieth century advances in the mathematical foundations of cryptography went hand-in-hand with technological developments. By using machines to scramble the plaintext very efficiently, it was hoped that it would be possible to achieve any desired level of security. Unfortunately, that hope was not realized. The capacity to devise complicated mechanisms for encryption and decryption always seemed to be matched by the capacity to devise mechanisms that can attack and break the resulting systems. The outcome was the conclusion that symmetric key cryptosystems can only provide security in a relative sense. The more complicated the procedures, the more difficult it is for an attack to be successful. But, at the same time, greater complication makes the system more difficult to use. Furthermore, systems can only be guaranteed secure against *known* methods of attack.

On the technological side, the most significant development has been the availability of electronic computers. These are ideal for calculations that involve iteration, and we shall describe a very useful procedure that takes advantage of this fact.

Suppose we are given a piece of text X, expressed as a string of n bits, and a key k, a string of s bits. Let F be a function that assigns to each such X and

k another string of n bits, $Y = F(k, X)$. In other words we have a function

$$F : \mathbb{F}_2{}^s \times \mathbb{F}_2{}^n \rightarrow \mathbb{F}_2{}^n.$$

For reasons of security, F should *not* be a 'nice' function, such as a linear function. When $s = 2$ and $n = 3$ we might use the function defined for a key $k = \alpha\beta$ and $X = x_1x_2x_3$ by the rule

$$y_1 = \alpha x_1 x_2 + \beta, \quad y_2 = x_2 + \beta x_3, \quad y_3 = (\alpha + \beta)x_1 x_3.$$

For a given key, this function is not linear, nor is it a bijection.

Definition 12.8 (Feistel iteration)

The *Feistel iteration* associated with F and the *initial values* $X_0, X_1 \in \mathbb{F}_2{}^n$, is defined as follows. Let $\mathbf{k} = (k_1, k_2, \ldots, k_r)$ be a sequence of keys, each of which is an element of of $\mathbb{F}_2{}^s$, and define

$$X_{i+1} = X_{i-1} + F(k_i, X_i) \qquad (i = 1, 2, \ldots, r),$$

where $+$ denotes the operation of bitwise addition in the vector space $\mathbb{F}_2{}^n$.

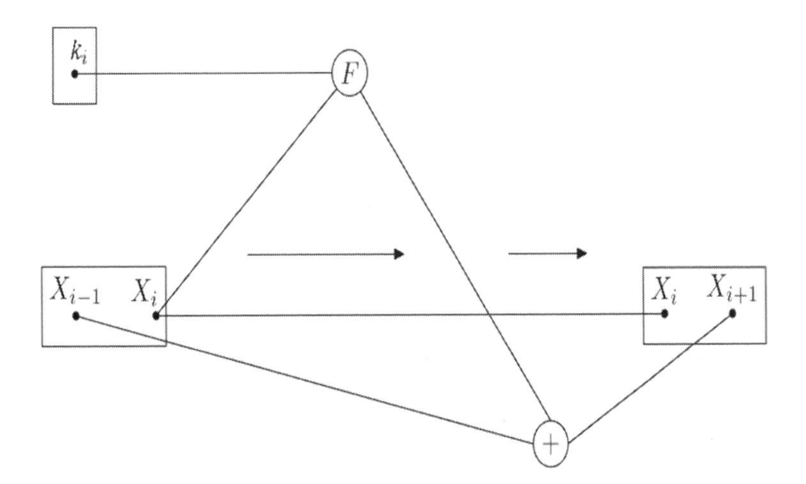

Figure 12.2 The ith round of a Feistel iteration

In cryptography each step of the iteration is called a 'round'. In the ith round the key k_i is used to transform the pair $X_{i-1}X_i$ into X_iX_{i+1} (Figure 12.2). Encryption is done by expressing all messages as sequences of strings of bits of length $2n$. If the plaintext contains a $2n$-string m, it is split into two equal parts, $m = X_0X_1$, and the n-strings X_0 and X_1 are taken as the initial values for the Feistel iteration. The corresponding ciphertext is the outcome after r rounds: $c = X_rX_{r+1}$. Thus the encryption function for a key-sequence $\mathbf{k}$ is given by

$$E_{\mathbf{k}}(m) = c, \quad \text{where } m = X_0X_1,\ c = X_rX_{r+1}.$$

Before we discuss the practical details, here is a simple example.

Example 12.9

Take $s = n = 3$, and define $Y = F(k, X)$ for a key $k = \alpha\beta\gamma$ by the rule

$$y_1y_2y_3 \;=\; F(\alpha\beta\gamma,\ x_1x_2x_3), \quad \text{where}$$

$$y_1 = \alpha x_1 + x_2x_3, \qquad y_2 = \beta x_2 + x_1x_3, \qquad y_3 = \gamma x_3 + x_1x_2.$$

If the key-sequence is $\mathbf{k} = (100, 101, 001, 111)$ and $m = 101\ 110$, calculate $c = E_{\mathbf{k}}(m)$.

Solution The calculation can be tabulated as follows.

i	k_i	X_i	$F(k_i, X_i)$	X_{i+1}
0	–	101	–	110
1	100	110	101	000
2	101	000	000	110
3	001	110	001	001
4	111	001	001	111

Thus $c = 001\ 111$.

The Feistel iteration provides unlimited scope for scrambling the data, since the parameters n, s, r can be as large as we please. Nevertheless it is trivially a symmetric key system!

Theorem 12.10

Suppose the message $m = X_0X_1$ is encrypted using the Feistel system with a function F and the key-sequence $\mathbf{k} = (k_1, k_2, \ldots, k_r)$, so that $c = E_{\mathbf{k}}(m) = X_rX_{r+1}$. Then the Feistel system with the same function F and key sequence $\mathbf{k}^* = (k_r, k_{r-1}, \ldots, k_1)$, when applied to $c' = X_{r+1}X_r$, yields $m' = X_1X_0$.

Proof

Since we are working in $\mathbb{F}_2{}^n$, the equation for the Feistel iteration can be rewritten as

$$X_{i-1} = X_{i+1} + F(k_i, X_i).$$

In this form the iteration proceeds in reverse: starting with $i = r$ the equation gives X_{r-1} in terms of X_{r+1} and X_r, and so on. Hence if the original key sequence is used in reverse, the same procedure can be used for decryption. □

We have already noted that the function F should not have 'nice' properties. For instance, if it is a linear function, the Feistel system will be vulnerable to a known plaintext attack (Exercise 12.12). As always, the security of the system will also depend on there being enough keys to guard against an attack by exhaustive search. All the keys $k_1, k_2, \ldots, k_r$ could be chosen independently, but in practice it is usual to 'extract' them by choosing specific sets of s bits from a master key K. For example, if 12 keys of length 32 are needed, Alice and Bob could agree on a master key of length 120, and take k_1 as bits 1 to 32, k_2 as bits 9 to 40, k_3 as bits 17 to 48, and so on.

EXERCISES

12.10. Take $s = 2$, $n = 3$, and define $Y = F(k, X)$ for a key $k = \alpha\beta$ by the rule $y_1y_2y_3 = F(\alpha\beta,\ x_1x_2x_3)$ where

$$y_1 = \beta(x_1 + x_2), \qquad y_2 = (\alpha + \beta)x_2, \qquad y_3 = \alpha + x_2x_3.$$

If the key-sequence is $\mathbf{k} = (11, 10, 10, 01)$ and $m = 011\,001$, calculate $c = E_{\mathbf{k}}(m)$.

12.11. Check your answer to the previous exercise by applying the relevant decryption function to c.

12.12. Suppose that, for any given key k, the function $X \mapsto F(k, X)$ is a linear function. Show that the entire Feistel iteration based on F is a linear function.

12.5 Encryption standards

In the 1970s the increasing use of cryptographic procedures in commerce and industry led to the suggestion that an 'encryption standard' should be established. In general, standards are used to communicate essential information to everyone, such as how to make nuts and bolts that will fit together. In cryptography, the aim is more subtle, because Alice and Bob must be able to communicate privately, without letting Eve know exactly how they are doing it.

Our discussion thus far has indicated that any cryptosystem must depend for its security on the secrecy of the key. Consequently, a system that purports to provide a standard of security must be prepared to answer two basic questions.

- Is the attack by exhaustive search feasible?
- Is there an attack that is better than exhaustive search?

In 1977 the first *Data Encryption Standard*, usually known as *DES*, was established. A central component is what is known as an *S-box*. In general terms, this is an array (matrix) S with rows corresponding to elements of ${\mathbb{F}_2}^M$, columns corresponding to elements of ${\mathbb{F}_2}^N$, and entries in ${\mathbb{F}_2}^R$. In DES there are 8 S-boxes, with parameters $M = 2$, $N = 4$, and $R = 4$. One of them is shown in Figure 12.3. For clarity, the rows are labelled $0-3$ as binary numbers, $0 = 00$, $1 = 01$, $2 = 10$, $3 = 11$. The column labels and the entries of S are denoted by $0 - 15$ in the same way: thus $5 = 0101$, $12 = 1100$, and so on.

	0	1	2	3	4	5	6	7	8	9	10	11	12	13	14	15
0	12	1	10	15	9	2	6	8	0	13	3	4	14	7	5	11
1	10	15	4	2	7	12	9	5	6	1	13	14	0	11	3	8
2	9	14	15	5	2	8	12	3	7	0	4	10	1	13	11	6
3	4	3	2	12	9	5	15	10	11	14	1	7	6	0	8	13

Figure 12.3 One of the S-boxes used in DES

Each S-box in DES determines a function $f_S : {\mathbb{F}_2}^6 \to {\mathbb{F}_2}^4$, defined by the rule

$$f_S(x_1x_2x_3x_4x_5x_6) \quad = \quad S(x_1x_6,\ x_2x_3x_4x_5).$$

For example, for the S-box given above,

$$f_S(111001) = S(11, 1100) = S(3, 12) = 6 = 0110.$$

The DES encryption function operates on blocks of 64 bits, using a key with (essentially) 56 bits. It is obtained by permuting the bits and combining the 8 functions f_S defined by the S-boxes in a complicated way, involving a Feistel iteration with 16 rounds. The details can be found in the references cited at the end of the chapter.

The S-boxes were constructed so that they satisfied a number of criteria, designed to make the system safe against known methods of attack. Two of these criteria were

- C1: each row of S is a permutation of ${\mathbb{F}_2}^4$;
- C2: if $x, y \in {\mathbb{F}_2}^6$ are such that the Hamming distance $d(x, y) = 1$, and $x' = f_S(x)$, $y' = f_S(y)$, then $d(x', y') \geq 2$.

The proposal for DES led to much controversy. The method of constructing the S-boxes was not revealed, a fact that some people regarded as suspicious. More seriously, the parameters chosen were thought to be too small, mainly because the master key had effectively only 56 bits. This meant that an attack by exhaustive search was not out of the question.

In fact, by the 1990s it was clear that DES could not answer either of the basic questions satisfactorily. Consequently it was replaced by the *Advanced Encryption Standard* or *AES*. In this system the master key has 128 bits, and it is currently not feasible to carry out an attack by exhaustive search. The system is not based directly on a Feistel iteration, but similar principles are applied. There is only one S-box, with $M = N = 4$ and $R = 8$ and it is defined by an explicit mathematical formula. So, if there are any 'hidden' features, everyone has an equal chance of discovering them. It will be interesting to see how long AES can survive.

EXERCISES

12.13. Let S be the S-box shown in Figure 12.3. Determine the values of $f_S(x)$ and $f_S(y)$ when $x = 101100$, and $y = 101101$, and verify that condition C2 holds in this case.

12.14. Let S be an S-box with $M = N = 2$ and $R = 3$, and let $f_S(x_1x_2x_3x_4) = S(x_1x_4,\ x_2x_3)$. Show that the appropriate form of condition C2 implies that the four entries in any row or column of S must be a code in ${\mathbb{F}_2}^3$ with minimum distance 2. Hence construct a suitable S.

12.6 The key distribution problem

In a symmetric key cryptosystem, Alice and Bob use similar keys. Metaphorically speaking, Alice puts the message in a box which she locks using the key

k. When Bob receives the box he unlocks it, using a key ℓ that is closely related to k. Writing $\ell = k'$ to emphasize this relationship, the procedure is illustrated in Figure 12.4.

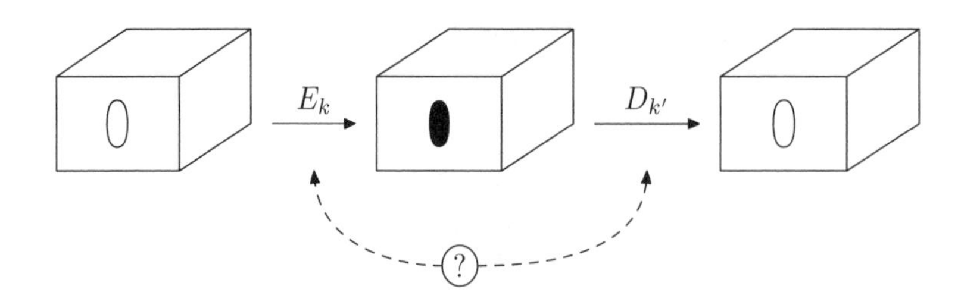

Figure 12.4 The symmetric key procedure

A major problem is that Alice and Bob must begin by communicating with each other in order to agree on the key, and this communication cannot therefore be protected by the encryption process. This fundamental difficulty is known as the *key distribution problem.*

There are several ways in which the key distribution problem might be overcome. If enormous resources are available, the one-time pad system could be used. In other words, before each message is transmitted Alice conveys a new key to Bob by some highly secure means, such as a courier protected by armed guards. Clearly, this is impractical for most purposes. A more practical approach is to consider alternatives to the symmetric key procedure.

> *Question: Is it possible to design a procedure in which Alice's key and Bob's key are independent, so that neither of them needs to know the other's key?*

At first sight, the following procedure appears to work.

1. Alice sends Bob a message encrypted using her key a.
2. Bob further encrypts using his key b and returns the message to Alice.
3. Alice decrypts using a' and sends the message to Bob.
4. Bob decrypts using b'.

This *double-locking* procedure is illustrated in Figure 12.5.

Unfortunately, our use of the metaphor about locking and unlocking boxes has concealed a serious difficulty. Denote the respective encryption and decryption functions by $E_a, D_{a'}, E_b, D_{b'}$, where $D_{a'}$ is the left-inverse of E_a and $D_{b'}$ is the left-inverse of E_b. Then the process applied to a message m is represented by the transformation

$$m \mapsto D_{b'}D_{a'}E_bE_a(m).$$

If it happens that $D_{a'}E_b = E_bD_{a'}$, this transformation will reduce to the identity, since in that case $D_{b'}D_{a'}E_bE_a(m) = D_{b'}E_bD_{a'}E_a(m) = m$. Here we use

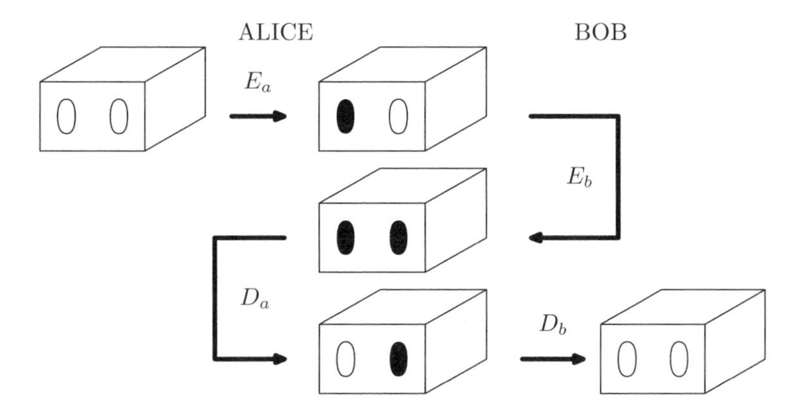

Figure 12.5 The double-locking procedure

the fact that, by definition, both $D_{a'}E_a$ and $D_{b'}E_{b'}$ are the identity. In other words, the double-locking procedure will work if Alice's operations commute with Bob's. It appears to work in the diagram because in practical terms the operations of turning keys in separate locks do commute.

Unfortunately, mathematical operations are not necessarily so well-behaved. For example, suppose the encryption and decryption keys are permutations of 26 letters, as in Section 10.5. Then it is easy to construct permutations that do not commute (see Exercise 12.15). Note that if Alice and Bob agree to restrict themselves to a set of permutations that do commute, then they can use the double-locking procedure. For example, they could use the Caesar system, in which each key is a permutation of the form α^k, where α is the permutation written in cycle notation as $\alpha = (\mathtt{ABCD} \cdots \mathtt{XYZ})$. In that case we have a set of 26 commuting keys α^k $(1 \leq k \leq 26)$. But this number is far too small to be useful in practice, and there remains the problem of how Alice and Bob can agree on which key to use. In the next chapter we shall describe a procedure that avoids the key distribution problem in a entirely different way.

EXERCISES

12.15. In Section 10.5 we described how permutations can be defined in terms of keywords. Alice has chosen the keyword `DEMOCRAT` and Bob has chosen `REPUBLICAN`. Write down the corresponding permutations δ and ρ in cycle notation, and show that the double-locking system fails in this case.

Further reading for Chapter 12

Perhaps the first advance with implications for modern cryptography was the introduction of the one-time pad. The original version was a practical system suggested by Vernam in 1917, and later patented. Soon afterwards Mauborgne observed that if the key is chosen uniformly at random then the system will have good security. In 1949 Shannon applied his seminal ideas to the science of cryptography, and gave a mathematical proof of this fact [**12.4**]. The idea was well known by the time of World War II, and a version of it was employed in the Lorenz machine used by the German high command. In 1941 an incompetent operator sent two slightly different messages with the same key (see Exercises 12.8 and 12.9), and this contributed significantly to the breaking of the system by Tutte and others at Bletchley Park, the British Cryptographic HQ [**12.7**]. In the 1960s the one-time pad (with suitably draconian safeguards) was used to transmit messages between Moscow and Washington. There are several interesting websites devoted to this topic.

Comprehensive accounts of current practice in cryptography, with details of DES, AES, and many topics not mentioned here, can be found in the books by Menezes *et al.* [**12.3**], Stinson [**12.5**], and Trappe and Washington [**12.6**]. The official specifications for DES [**12.2**] and AES [**12.1**] are also worth reading.

12.1 *Advanced Encryption Standard (AES).* Federal Information Processing Standard (FIPS), Publication 197 (2001).

12.2 *Data Encryption Standard (DES).* Federal Information Processing Standard (FIPS), Publication 46 (1977).

12.3 A.J. Menezes, P.C. van Oorschott, S.A. Vanstone. *Handbook of Applied Cryptography.* CRC Press (1996).

12.4 C. Shannon. Communication theory of secrecy systems. *Bell Syst. Tech. J.* 28 (1949) 657-715.

12.5 D.R. Stinson. *Cryptography, Theory and Practice.* Chapman and Hall (third edition, 2006).

12.6 W. Trappe and L.C. Washington. *Introduction to Cryptography and Coding Theory.* Prentice-Hall (2003).

12.7 W.T. Tutte. Fish and I. In: *Coding Theory and Cryptology* (D. Joyner ed.). Springer, New York (2000).

13 The RSA cryptosystem

13.1 A new approach to cryptography

In the previous chapter we noted that symmetric key cryptosystems are limited by the problem of distributing keys. A radical new approach was developed in the 1970s, known as *public key cryptography.*

The fundamental idea is that a typical user (Bob) has two keys, a *public key* and a *private key.* The public key is used by Alice and others to encrypt messages that they wish to send to Bob, and the private key is used by Bob to decrypt these messages. The security of the system depends on ensuring that Bob's private key cannot be found easily, even though everyone knows his public key.

A practical method of implementing this idea was discovered by Rivest, Shamir, and Adleman in 1977 and is known as the *RSA* cryptosystem. There are now many other systems of public key cryptography, but RSA is still widely used, and it is worth studying because it illustrates the basic principles simply and elegantly.

In the RSA system the plaintexts are sequences of *integers mod* n, where it is reasonable to think of n as a number with at least 300 decimal digits. Any simple rule can be used to convert the 'raw' message to plaintext. For instance, if the message is written in the 27-symbol alphabet $\mathbb{A}$, we could begin by representing each symbol as a 5-bit binary number:

$$\sqcup \mapsto 00000, \quad \mathtt{A} \mapsto 00001, \quad \mathtt{B} \mapsto 00010, \quad \ldots, \quad \mathtt{Z} \mapsto 11010.$$

N. L. Biggs, *An Introduction to Information Communication and Cryptography*,
DOI: 10.1007/978-1-84800-273-9_13, © Springer-Verlag London Limited 2008

Then we could form blocks of (for example) 180 symbols, making strings of 900 binary digits, which represent numbers with about 300 decimal digits. These numbers can then be treated as integers mod n, provided n is large enough. In practice, a system based on the ASCII or Unicode standards would be used: the point is that the plaintext can be constructed, and converted back into its 'raw' form, using publicly available information.

We have already observed on several occasions that the set of integers mod n, with the operations of addition and multiplication, is a *ring*. We denote it by $\mathbb{Z}_n$. In general, not every non-zero element of $\mathbb{Z}_n$ has a multiplicative inverse. We assume that the reader is familiar with the fact that a non-zero element $x \in \mathbb{Z}_n$ has a multiplicative inverse $x^{-1} \in \mathbb{Z}_n$ if and only if $\gcd(x, n) = 1$. In Section 13.3 we shall describe a good method of calculating x^{-1}.

Definition 13.1 (ϕ function)

For any positive integer n, the number of integers x in the range $1 \le x \le n$ such that $\gcd(x, n) = 1$ is denoted by $\phi(n)$. It follows from the result stated above that $\phi(n)$ is also the number of invertible elements of $\mathbb{Z}_n$.

For example, taking $n = 14$ the values of x such that $\gcd(x, 14) = 1$ are $1, 3, 5, 9, 11, 13$ and so $\phi(14) = 6$. In this case the inverses are easily found by inspection: thus $3^{-1} = 5$ since $3 \times 5 = 1 \pmod{14}$, and so on.

If n is a prime number, say $n = p$, then the $p-1$ numbers $x = 1, 2, \ldots, p-1$ all satisfy $\gcd(x, p) = 1$, and so $\phi(p) = p-1$. We shall need the following simple extension of this rule.

Lemma 13.2

If $n = pq$, where p and q are primes, then $\phi(n) = (p-1)(q-1)$.

Proof

The integers x such that $\gcd(x, n) \ne 1$ are the multiples of p and the multiples of q. In the range $1 \le x \le n$ these are:

$$\begin{array}{llllll} p & 2p & 3p & \ldots & (q-1)p & qp \qquad \text{(there are } q \text{ of these);} \\ q & 2q & 3q & \ldots & (p-1)q & pq \qquad \text{(there are } p \text{ of these).} \end{array}$$

Since $qp = pq = n$, the total number is $p + q - 1$, so we have

$$\phi(n) = n - (p + q - 1) = pq - p - q + 1 = (p-1)(q-1).$$

$\square$

EXERCISES

13.1. Make a list of those elements of $\mathbb{Z}_{36}$ that have multiplicative inverses, and find the inverse of 5 mod 36.

13.2. Evaluate $\phi(257)$ and $\phi(253)$.

13.3. Given that x has the inverse y mod n, find the inverse of $n - x$ mod n. Deduce that $\phi(n)$ is an even number, except when $n = 1$ and $n = 2$.

13.2 Outline of the RSA system

In the RSA system there are a number of users, including, as always, Alice and Bob. Each user, say Bob, has an encryption function and a decryption function, constructed according to the following rules.

- Choose two prime numbers p, q and calculate

$$n = pq, \quad \phi = (p-1)(q-1).$$

- Choose e such that $\gcd(e, \phi) = 1$, and calculate

$$d = e^{-1} \pmod{\phi}.$$

The encryption and decryption functions are defined as follows:

$$E_{n,e}(m) = m^e \qquad (m \in \mathbb{Z}_n),$$

$$D_{n,d}(c) = c^d \qquad (c \in \mathbb{Z}_n).$$

The system works in the following way. Starting with p and q, Bob uses the rules given above to construct, in turn, the numbers n, ϕ, e, and d. He makes his *public key* (n, e) available to everyone, but keeps his *private key* d secret. When Alice wishes to send Bob a message, she expresses it in the form of a sequence of integers m mod n, calculates $c = E_{n,e}(m)$, and sends c. Bob then uses his private key to compute $D_{n,d}(c)$. (Note that n is not private, but it is needed in the construction of $D_{n,d}$.)

The following example uses very small numbers and is presented for the purposes of illustration only. The calculations can be checked by hand (although some hard work is needed in part (iii)). The information that Bob must keep private is in **bold** type.

Example 13.3

Suppose Bob has chosen $p = \mathbf{47}, q = \mathbf{59}$.

(i) Find n and ϕ, and show that $e = 157$ is a valid choice for his public key.
(ii) Verify that his private key is $d = \mathbf{17}$.
(iii) How does Alice send Bob a message represented by a sequence of integers mod n, containing (for example) the integer $m = 5$? How does Bob decrypt it?

Solution (i) We have

$$n = \mathbf{47} \times \mathbf{59} = 2773, \quad \phi = \mathbf{46} \times \mathbf{58} = \mathbf{2668}.$$

The choice $e = 157$ is valid provided that $\gcd(2668, 157) = 1$. This condition holds because 157 is a prime and does not divide 2668.

(ii) We can check this by calculating

$$de = 17 \times 157 = 2669 = 1 \pmod{2668}.$$

(iii) If Alice wishes to send Bob a message, she looks up his public key $(n, e) = (2773, 157)$, and converts her message into a sequence of integers mod n. She encrypts using the encryption function $E_{n,e}(m) = m^e$, and sends the ciphertext to Bob. In this case

$$c = 5^{157} = 1044 \pmod{2773}.$$

When Bob receives the ciphertext $c = 1044$ he applies his (private) decryption function $D_{n,d}(c) = c^d$, obtaining

$$m' = 1044^{17} = 5 \pmod{2773}.$$

Since $m' = m$, the original message is recovered.

In reality, the numbers must be much larger, and Bob would require the help of a computer algebra system, such as MAPLE, to set up as a user. To find the two primes p and q, he could use the MAPLE command `nextprime(x)`, which returns the next prime number greater than x. For instance, the command

```
>p:= nextprime(13^151); q:= nextprime(19^132);
```

will result in the output

$$\begin{aligned} p = &\quad 160489\cdots \quad \cdots \quad \cdots 323547 \\ q = &\quad 624417\cdots \quad \cdots \quad \cdots 576463 \end{aligned}$$

where there are over 150 digits in each number. The significant fact is that the numbers p and q are found almost instantaneously, so this computation can be said to be 'easy'. The reason is far from obvious: it is a consequence of the existence of good algorithms for *primality testing*, the details of which can be found in the references given at the end of the chapter.

It is rather less surprising that, with the aid of MAPLE, Bob can easily calculate $n = pq$:

```
>n:= p*q;
```

$$n = 100212\cdots \quad \cdots \quad \cdots \quad \cdots 874261$$

where n has over 300 digits.

In order to justify the usefulness of RSA as a practical cryptosystem, it remains to provide satisfactory answers to three questions.

- *Feasibility* Why is it easy to carry out the calculations involved, such as finding $d = e^{-1} \pmod{\phi}$, and computing the values of m^e and $c^d \pmod{n}$?
- *Correctness* Why is the function $D_{n,d}$ the left-inverse of $E_{n,e}$? That is, why is it true that
$$D_{n,d}(E_{n,e}(m)) = m \quad \text{for all } m \in \mathbb{Z}_n?$$
- *Confidentiality* Why cannot another user (Eve) discover the plaintext m, given that she knows the public key (n, e) and the ciphertext $c = E_{n,e}(m)$?

These questions will be answered in the next three sections, on the assumption that the number $n = pq$ is 'large'. The following exercises illustrate the fact that the system is vulnerable if n small.

EXERCISES

13.4. George and Tony used the RSA cryptosystem. George advertised his public key $n = 77$, $e = 43$. What were the primes p and q, and what was his private key? Tony sent him a message $m \in \mathbb{Z}_n$, and Vladimir intercepted the ciphertext $c = 5$. What was m?

13.5. Tony advertised the public key $n = 3599$, $e = 31$. Unfortunately, he has become confused as to whether his PIN (private key) is 3301, 3031, or 3013. Which is it?

13.6. A naive user of RSA has announced the public key $n = 2903239$, $e = 5$. Eve has worked out that $n = 1237 \times 2347$. Verify that the public key is valid and explain why the private key is $d = 2319725$. [Calculators are not required.]

13.3 Feasibility of RSA

We shall now explain why the calculations involved in RSA can be done easily. This depends on the existence of two algorithms, both of which are used to reduce complicated problems to basic arithmetic.

The *Euclidean algorithm* is essentially a method for calculating the greatest common divisor $\gcd(a, b)$ of two integers a and b (we can assume $a < b$). It depends on the fact that if $b = qa + r$ then $\gcd(a, b) = \gcd(r, a)$. We can therefore replace (a, b) by $(a', b') = (r, a)$, and repeat the process. Eventally we obtain a pair (a^*, b^*) in which a^* is a divisor of b^*.

Example 13.4

Find the greatest common divisor of 654 and 2406.

Solution The calculation proceeds as follows:

$$\begin{aligned} 2406 &= 3 \times 654 \quad +444 \\ 654 &= 1 \times 444 \quad +210 \\ 444 &= 2 \times 210 \quad +24 \\ 210 &= 8 \times 24 \quad +18 \\ 24 &= 1 \times 18 \quad +6 \\ 18 &= 3 \times 6. \end{aligned}$$

Consequently 6 is the greatest number that divides $18, 24, 210, 444, 654, 2406$, and in particular $\gcd(654, 2406) = 6$.

If $\gcd(a, b) = 1$, then a has a multiplicative inverse mod b, and the Euclidean algorithm can be used to calculate it. This is done by reversing the calculation, so that 1 is expressed in the form $\lambda a + \mu b$, where λ and μ are integers. The equation $1 = \lambda a + \mu b$ can be written in the form $\lambda a = 1 - \mu b$, that is, $\lambda a = 1 \pmod{b}$. Thus λ is the inverse of $a \pmod{b}$.

Example 13.5

Find the inverse of 24 mod 31.

Solution First we check that $\gcd(24, 31) = 1$:

$$\begin{aligned} 31 &= 1 \times 24 \quad +7 \\ 24 &= 3 \times 7 \quad +3 \\ 7 &= 2 \times 3 \quad +1 \\ 3 &= 3 \times 1. \end{aligned}$$

Now, starting with the last-but-one equation and working backwards, we have

$$\begin{aligned} 1 &= 7 - 2 \times 3 \\ &= 7 - 2 \times (24 - (3 \times 7)) &&= 7 \times 7 - 2 \times 24 \\ &= 7 \times (31 - 1 \times 24) - 2 \times 24 &&= -9 \times 24 + 7 \times 31. \end{aligned}$$

Hence $(-9) \times 24 = 1 \pmod{31}$, so $24^{-1} = -9 = 22$.

Example 13.6

In Example 13.3 Bob set up as a user of RSA by choosing $p = 47$, $q = 59$, so that $n = 2773$, $\phi = 2668$. Use the Euclidean algorithm to show that $e = 157$ is a valid choice for his public key, and explain how he can calculate his private key $d = e^{-1} \pmod{\phi}$.

Solution The Euclidean algorithm shows that $\gcd(e, \phi) = 1$:

$$\begin{aligned} 2668 &= 16 \times 157 \quad +156 \\ 157 &= 1 \times 156 \quad + 1 \\ 156 &= 156 \times 1\,. \end{aligned}$$

$d = e^{-1}$ is found by working backwards:

$$\begin{aligned} 1 &= 157 - 1 \times 156 \\ &= 157 - 1 \times (2668 - (16 \times 157)) \\ &= (-1) \times 2668 + 17 \times 157. \end{aligned}$$

Thus $17 \times 157 = 1 \pmod{2668}$, so $d = 17$ is Bob's private key.

In the real world the numbers are huge, and a computer must be used. It can perform the calculations very quickly. For example, in MAPLE the command

```
>d:= e^(-1) mod phi;
```

will return the value of d if e has an inverse mod ϕ, or an error message if not.

The second useful algorithm is *repeated-squaring*, which is used to calculate the power b^k, where b and k are given positive integers. In RSA the processes of encryption and decryption both involve calculation of this kind, since the plaintext m and the ciphertext c are related by the equations $c = m^e$, $m = c^d$.

Naively, it seems that in order to compute the power b^k we must use $k - 1$ multiplications, since $b^k = b \times b \times \cdots \times b$. However the repeated squaring algorithm provides a much better way.

Example 13.7

Suppose the number b is given. How many multiplications are needed to compute b^{16} and how many more multiplications are needed to compute b^{23}?

Solution The calculation of b^{16} requires only four multiplications:

$$b^2 = b \times b, \quad b^4 = b^2 \times b^2, \quad b^8 = b^4 \times b^4, \quad b^{16} = b^8 \times b^8.$$

Given this calculation, we can use the fact that $23 = 16 + 4 + 2 + 1$, so that only three more multiplications are needed to find b^{23}:

$$b^{23} = b^{16} \times b^4 \times b^2 \times b.$$

Consequently b^{23} can be found with only $4 + 3 = 7$ multiplications in all.

The method used in the example can be used to calculate any power b^k, as follows. The first stage is to calculate, in turn,

$$b^2,\ b^4,\ b^8,\ \ldots,\ b^{2^i},$$

where 2^i is the largest power of 2 not exceeding k. This requires i multiplications. At the second stage k is expressed as a sum of a subset of the numbers $1, 2, 4, \ldots, 2^i$, and the corresponding powers of b multiplied to give b^k. This stage requires at most i multiplications, and consequently the total number is at most $2i$.

Since $2^i \leq k$, the bound $2i$ is approximately $2\log_2 k$. This number is proportional to the number of digits required to represent k (rather than k itself), because writing k in the usual decimal representation requires $\log_{10} k$ digits. Since $\log_{10} k = \log_2 k \;/\log_2 10$, it follows that for a number with ℓ decimal digits, the number of multiplications is at most

$$(2\log_2 10) \times \ell \;\approx\; 6.64\,\ell.$$

For example, for a number with 150 digits, about 10^3 multiplications are needed. By comparison the naive method would require 10^{150}.

In the application to RSA there is the further complication that multiplication must be carried out modulo an integer n. After each multiplication, the remainder on division by n must be found. In MAPLE this operation can be implemented by the single command

```
>b &^ k (mod n);
```

For future reference it is worth noting that, although these calculations are feasible, they cannot be said to be trivial. We shall return to this topic in Section 15.5.

EXERCISES

13.7. Use the Euclidean algorithm to show that $\gcd(15,68) = 1$. Hence find the inverse of 15 mod 68.

13.8. Given a positive integer b, explain how to calculate b^{55} with only 9 multiplications, using the repeated-squaring algorithm.

13.9. Approximately how many multiplications are needed to calculate $b^{3578674567}$ by the repeated-squaring algorithm?

13.10. A naive user of RSA has announced the public key $n = 187$, $e = 23$. Find the factors of n and hence (i) verify that the public key is valid and (ii) determine the private key. If the user receives the integer 6 (mod 187) as a piece of ciphertext, what number m represents the plaintext? What function was used by the sender to encrypt m, and how many multiplications are required to calculate this function?

13.11. Calculate 2^{1000} mod 47. [MAPLE can be used, but it is possible to do the calculation 'by hand'.]

13.12. The repeated-squaring algorithm does not necessarily use the least possible number of multiplications. For example, there is a method of calculating b^{55} using only 8 multiplications, rather than the 9 used in Exercise 13.8. Find this method. [See also Section 15.5.]

13.4 Correctness of RSA

The next task is to explain why RSA works – that is, why the function $D_{n,d}$ is the left-inverse of $E_{n,e}$. According to the definitions, the condition

$$D_{n,d}(E_{n,e}(m)) = m$$

reduces to $(m^e)^d = m \pmod{n}$, so we have to prove that this holds when d is the inverse of e mod $\phi(n)$.

Lemma 13.8

If $\gcd(x,n) = 1$, so that x has an inverse in $\mathbb{Z}_n$, then $x^{\phi(n)} = 1$ in $\mathbb{Z}_n$.

Proof

Suppose the invertible elements of $\mathbb{Z}_n$ are $b_1, b_2, \ldots, b_f$, where we write $f =$

$\phi(n)$ for short. Given that $\gcd(x, n) = 1$, x is invertible, and each xb_i is also invertible, since the inverse of xb_i is $b_i^{-1}x^{-1}$. Thus the sets $\{b_1, b_2, \ldots, b_f\}$ and $\{xb_1, xb_2, \ldots, xb_f\}$ have the same members. Since multiplication in $\mathbb{Z}_n$ is commutative the product of the elements is the same, whatever the order, and so

$$b_1 b_2 \cdots b_f = (xb_1)(xb_2) \cdots (xb_f) = x^f (b_1 b_2 \cdots b_f) \quad \text{in } \mathbb{Z}_n.$$

Hence $x^f = 1$, and since $f = \phi(n)$ the result follows. □

In the following proof we assume that the plaintext element $m \in \mathbb{Z}_n$ is such that $\gcd(m, n) = 1$. In practice $n = pq$, where p and q are primes with at least 150 digits, so the probability that an element of $\mathbb{Z}_n$ does *not* satisfy this condition is $(p + q - 1)/pq$ (Lemma 13.2). This is of the order of 10^{-150}, so the probability is infinitesimally small. (In fact the result does hold when $\gcd(m, n) \neq 1$, but a slightly different proof is needed – see Exercise 13.15.)

Theorem 13.9

If the RSA encryption and decryption functions are $E_{n,e}$ and $D_{n,d}$, and $\gcd(m, n) = 1$, then

$$D_{n,d}(E_{n,e}(m)) = m.$$

That is, $m^{ed} = m \pmod{n}$.

Proof

In the RSA construction, starting with p, q, Bob calculates $n = pq$ and $\phi = (p - 1)(q - 1)$, so $\phi = \phi(n)$, by Lemma 13.2. Bob then chooses e so that $\gcd(e, \phi) = 1$, so that e has an inverse d mod ϕ. Explicitly

$$ed = 1 + \phi\alpha, \quad \text{for some integer } \alpha.$$

It follows from Lemma 13.8 that

$$m^{ed} = m^{1+\phi\alpha} = m \times (m^{\phi})^{\alpha} = m \pmod{n},$$

as claimed. □

EXERCISES

13.13. If $n = 15$ and $e = 3$, what is d? Verify explicitly that in this case $D_{n,d}(E_{n,e}(m)) = m$ for all $m \in \mathbb{Z}_n$, including those values of m for which $\gcd(m, n) \neq 1$.

13.14. Show that the statement $m^{\phi(n)} = 1 \pmod{n}$ does not hold for all m in the range $1 \leq m < n$ when $n = 15$.

13.15. Let n be the product of primes p and q. Let m be such that $\gcd(m, n) \neq 1$ and $1 \leq m < pq$, and let K be a multiple of $\phi(n)$. Show that

$$m^{K+1} = m \pmod{p}, \qquad m^{K+1} = m \pmod{q}.$$

Deduce that the RSA encryption-decryption process works even when $\gcd(m, n) \neq 1$.

13.5 Confidentiality of RSA

It remains to explain why Eve cannot discover the plaintext m, given that she knows the public key (n, e) and may obtain the ciphertext $c = E_{n,e}(m)$.

By definition $m = D_{n,d}(c) = c^d \pmod{n}$. In order to calculate m, it would be enough for Eve to know the private key d, which requires knowledge of ϕ, since d is the inverse of e mod ϕ. This in turn requires knowledge of the primes p and q. But Eve knows only n, not the factorization $n = pq$. Although there is a good algorithm for multiplying p and q (long multiplication, as taught in elementary arithmetic) no one has yet found a good algorithm for 'unmultiplying': that is, for finding p and q when n is given. (It is worth noting that, as a consequence of the 'Fundamental Theorem of Arithmetic', p and q are unique.) For example, if Eve has access to MAPLE, she could attempt to find the factors p and q by using the command

```
>ifactor(n);
```

which will work if n is small (see Exercise 13.16). But if n has 300 decimal digits the algorithms currently available (2008) will not respond to Eve's command, even if she is prepared to wait for a lifetime. This suggests that the problem is 'hard'. Thus, for the time being, Eve cannot expect to break RSA by factorizing n. The possibility of a successful attack by a different method remains open.

In summary, the 'secret components' for each user are the pair (p, q), and the values of ϕ and d. In fact, if Eve obtains any one of these three quantities, she can calculate the others, and the security of the system for that user will be destroyed. For example, if Eve knows ϕ, then not only can she calculate d, she can also calculate p and q, as in the following example.

Example 13.10

Bob is using RSA with public key $n = 28199$, $e = 137$. Eve has discovered that Bob's value of ϕ is 27864. Find the primes p, q such that $n = pq$.

Solution We have

$$pq = 28199, \quad (p-1)(q-1) = 27864,$$

so that $p+q = 336$. Given the sum and product of p and q, it follows that they are the roots of the quadratic equation

$$x^2 - 336x + 28199 = 0.$$

Solving this equation by factorizing the quadratic expression is no easier than factorizing 28199. But fortunately we know a formula that reduces the problem to the calculation of a square root:

$$p, q = \frac{1}{2}\left(336 \pm \sqrt{336^2 - 4 \times 28199}\right).$$

There is a good algorithm for calculating square roots, although it is not really needed in this example, since

$$336^2 - 4 \times 28199 = 112896 - 112796 = 100.$$

So the roots are $(336 \pm 10)/2$, that is $p = 173$, $q = 163$.

We have given plausible reasons why RSA can be implemented as a working cryptosystem. It is worth repeating that the most significant property, *confidentiality*, is not a mathematically proven fact. It relies on the presumed difficulty of factoring integers, and if there is a significant advance in that direction, the system might be compromised.

EXERCISES

13.16. The Department of Bureaucracy has advertised that it will use the RSA cryptosystem with the public key $n = 173113$, $e = 7$. Using the following MAPLE output, show that the Department's choice of public key is valid, and find the private key.

```
>ifactor(173113);
```

$$(331)(523)$$

13.17. Bob is using RSA with public key $n = 4189$, $e = 97$. Eve has discovered that Bob's value of ϕ is 4060. Find the primes p, q such that $n = pq$ and Bob's private key d.

13.18. Chang is using RSA with public key

$$n = 247\,064\,529\,085\,306\,223\,003\,563,$$

$$e = 145\,268\,762\,498\,836\,504\,194\,121.$$

Deng has discovered that the private key $d = 987654321$ will successfully decrypt messages addressed to Chang. Explain how, *without factorizing* n, Deng can also obtain ϕ and the primes p, q used by Chang to construct the keys.

13.19. A user of the RSA system has mistakenly chosen a public key (n, e) with n prime. Show that messages sent to this user can be decrypted easily.

13.20. Although the general problem of factorizing a number with (say) 500 digits is still thought to be intractable, there are some numbers with 500 digits for which some factors can be found relatively easily. For example, consider the number

$$N = 12345678901234567890 \cdots 1234567890,$$

where there are 50 blocks of the digits 1234567890. Can you find factors of N? Can you find the prime factors of N? Can you find the prime factors of $N + 1$?

Further reading for Chapter 13

The idea of public-key cryptography was first suggested by Diffie and Hellman [**13.4**] in 1976. About a year later Rivest, Shamir, and Adleman proposed a practical system, and it was published by Martin Gardner in the *Scientific American* [**13.5**].

In fact a system very similar to RSA had been suggested a few years earlier by three mathematicians, Cocks, Ellis, and Williamson, working for the British cryptographic service (GCHQ). But this fact remained secret and was not made public until 1997: for the details see Singh [**1.3**, 279-292].

The discovery of RSA stimulated a great deal of research into the computational aspects of number theory. A standard reference is the book by Crandall and Pomerance [**13.3**]. From the RSA perspective there are two major questions. The first is the *factoring* problem, the difficulty of which is the basis for the belief that the system is secure. As an illustration, Gardner [**13.5**] challenged readers of the *Scientific American* to factorize a number with 129 digits. This number was not factorized until 1994, by the combined efforts of over 600 people [**13.2**]. The factoring problem is still believed to be 'hard', but there is as yet no mathematical proof of this statement. The second problem is *primality testing*, which is needed to construct the primes p and q. For this problem probabilistic methods that work well in practice, such as the Solovay-Strassen algorithm [**13.9**] and the Miller-Rabin algorithm [**13.7**, **13.8**], have been known since the 1970s. More recently Agarwal, Kayal, and Saxena [**13.1**] succeeded in constructing a deterministic algorithm that is technically 'good', but the probabilistic methods are more effective in practice.

There are numerous attacks on RSA that use special features of the numbers involved. Also, the traditional ingenuity of codebreakers has also resulted in suggestions for attacks based on quite different considerations – analogous to the attack on a substitution system by frequency analysis. Such methods are known as *side-channel attacks*. One of the first was discovered by Kocher [**13.6**], then a student at Stanford, who noticed that the inner workings of the repeated-squaring algorithm could be revealed by recording how much time was needed to do the calculations.

13.1 M. Agarwal, N. Kayal, N. Saxena. PRIMES is in P. *Annals of Math.* 160 (2004) 781-793.

13.2 D. Atkins, M.Graff, A. Lenstra, P. Leyland. The magic words are squeamish ossifrage. *Advances in Cryptology - ASIACRYPT '94* (1995) 703-722.

13.3 R. Crandall and C. Pomerance. *Prime Numbers: A Computational Perspective.* Springer-Telos (2000).

13.4 W. Diffie and M. Hellman. New directions in cryptography. *IEEE Trans. in Information Theory* 22 (1976) 644-654.

13.5 M. Gardner. A new kind of cipher that would take millions of years to break. *Scientific American* 237 (August 1977) 120-124.

13.6 P. Kocher. Timing attacks on implementations of Diffie-Hellman, RSA, DSS and other systems. *Advances in Cryptology - CRYPTO '96* (1996) 104-113.

13.7 G.L. Miller. Riemann's hypothesis and tests for primality. *Journal of Computer and Systems Science* 13 (1976) 300-317.

13.8 M.O. Rabin. Probabilistic algorithms for testing primality. *Journal of Number Theory* 12 (1980) 128-138.

13.9 R. Solovay and V. Strassen. A fast Monte Carlo test for primality. *SIAM Journal on Computing* 6 (1977) 84-85.

14
Cryptography and calculation

14.1 The scope of cryptography

Modern cryptography is not just about sending secret messages. It covers many aspects of security, including authentication, integrity, and non-repudiation.

- *Authentication* Bob must be sure that a message that purports to come from Alice really does come from her.
- *Integrity* Eve should not be able to alter a message from Alice to Bob.
- *Non-repudiation* Alice should not be able to claim that she was not responsible for a message that she has sent.

In this chapter we shall consider how the 'public key' approach can be applied to these topics.

The general framework for a public key cryptosystem is a simple abstraction from the RSA system discussed in the previous chapter. There are sets $\mathcal{M}$, $\mathcal{C}$, and $\mathcal{K}$, of plaintext messages, ciphertext messages and keys, respectively, and a typical user (Alice) has two keys: a public key a and a private key a'. These keys determine an encryption function E_a and its left-inverse, a decryption function $D_{a'}$:

$$D_{a'}(E_a(m)) = m \quad \text{for all } m \in \mathcal{M}.$$

The security of the system depends on the fact that, given a, it is hard to find a'.

Unfortunately, the precise definition of the words 'easy' and 'hard' is rather complicated. The branch of theoretical computer science known as *complexity*

N. L. Biggs, *An Introduction to Information Communication and Cryptography*,
DOI: 10.1007/978-1-84800-273-9_14,

theory is concerned with making mathematically sound statements about 'easy' and 'hard' computations, but the theory is still being developed. For expository purposes it is enough to rely on practical experience of computer algebra systems such as MAPLE. In Chapter 13 we used this approach to back up the claim that RSA is a practicable system, based on the fact that multiplying is easy, but unmultiplying is hard. In Section 14.4 we shall describe another operation that is easy to do, but hard to undo. As we shall see, it forms the basis of many security procedures used in modern cryptography.

14.2 Hashing

A simple device that is often useful in cryptography is known as *hashing*. The idea is that a message, which may be of any length, is reduced to a 'message digest', which is a string of bits with a fixed length. We shall explain how this process can be used to authenticate a message, and to guarantee its integrity – with some provisos.

Definition 14.1 (Hash function)

Let n be fixed positive integer (in practice, the value $n = 160$ is often used). A *hash function* for a set of messages $\mathcal{M}$ is a function $h : \mathcal{M} \to {\mathbb{F}_2}^n$.

A simple application of this construction is that when Alice wishes to send Bob a message m, she actually sends the pair $(m, h(m))$. This pair may or may not be encrypted by Alice and decrypted by Bob in the usual way; in either case Bob should obtain a pair (x, y) with $x = m$ and $y = h(m)$. If Bob gets a pair (x, y) such that $h(x) \neq y$, he knows that something is wrong. Note that it is quite possible that the 'false' message x is meaningful: it may have been sent by Eve, with the intention of causing trouble.

Since messages can have any length, we can assume that the size of $\mathcal{M}$ is greater than 2^n. This means that h cannot be an injective function: there will surely be two different messages m and m' such that $h(m) = h(m')$. If h is constructed simplistically, then it will be easy to find examples.

Example 14.2

Suppose a hash function h is constructed as follows. The message m is expressed as a string of bits, and split into blocks of length n (using some 0's to fill in the last block, if necessary). Then $h(m)$ defined to be the sum (in ${\mathbb{F}_2}^n$) of the blocks. Given $m \in \mathcal{M}$ explain how to construct $m' \in \mathcal{M}$ such that $m' \neq m$ and $h(m') = h(m)$.

Solution There are many ways in which this can be achieved. If m' is obtained by adding a specific string s of n bits to two blocks in m, then $s + s$ is the zero block and $h(m') = h(m)$. This procedure can be extended in obvious ways to produce a message that is apparently very different from m, but has the same $h(m)$.

In cryptography, a pair $m \neq m'$ such that $h(m) = h(m')$ is known as a *collision*. Clearly, in order to guarantee the authenticity and integrity of messages it is important that collisions are hard to find, even though they are bound to exist. A hash function that has these properties is said to be *collision-resistant*. Many such functions have been proposed, and some of them have been widely used. However, it is worth noting that collisions have been constructed in several of the most popular hash functions.

The hashing procedure can be strengthened significantly if Alice and Bob agree to use a family of hash functions $\{h_k\}$, where the key $k \in \mathcal{K}$ is secret.

Definition 14.3 (Message authentication code)

A *message authentication code (MAC)* is a family of hash functions h_k $(k \in \mathcal{K})$ such that

(i) given $m \in \mathcal{M}$ and $k \in \mathcal{K}$, it is easy to compute $h_k(m)$;

(ii) given a set of pairs $(m_i, h_k(m_i))$ $(1 \leq i \leq r)$, but not k, it is hard to find $h_k(m)$ for any m that is not one of the m_i.

The conditions are designed for the situation in which Eve's goal is to construct a false message m with a valid hash $h_k(m)$. If Eve is lucky she may be able to intercept any number of messages m_i and their hashes $h_k(m_i)$. Thus it is important that this information will not enable her construct $h_k(m)$.

We shall return to the topics of authentication, integrity, and non-repudiation in Section 14.7.

EXERCISES

14.1. Let $h : \mathcal{M} \to \mathbb{F}_2{}^n$ be a hash function, and define

$$\mu_y = |\{x \mid h(x) = y\}| \ \ (y \in \mathbb{F}_2{}^n), \quad \sigma = |\{\{x_1, x_2\} \mid h(x_1) = h(x_2)\}|.$$

Note that σ counts the collisions as the number of *unordered* pairs. If $M = |\mathcal{M}|$, show that

$$\sum_y \mu_y = M, \qquad \sigma = \frac{1}{2}\sum_y {\mu_y}^2 - \frac{M}{2}.$$

14.2. Deduce from the results of the previous exercise that the number of collisions is at least $\frac{M}{2N}(M - N)$, where $N = 2^n$.

14.3. Show that if $\{h_k\}$ $(k \in \mathcal{K})$ is a MAC then each h_k must be collision-resistant.

14.4. Suppose that $\mathcal{M}$ is a vector space and h is a linear transformation $\mathcal{M} \to {\mathbb{F}_2}^n$. Show that h does not satisfy condition (ii) of Definition 14.3.

14.3 Calculations in the field $\mathbb{F}_p$

We have already made use of the fact that when p is prime, the set of integers mod p, with the relevant arithmetical operations, is a *field*, denoted by $\mathbb{F}_p$. Specifically, this means that every element of $\mathbb{F}_p$ except 0 has a multiplicative inverse, and so the set of non-zero elements of $\mathbb{F}_p$ forms a group with the operation of multiplication. This group will be denoted by $\mathbb{F}_p^\times$, and its members by

$$1, 2, \ \ldots \ , p-1$$

in the usual way.

It is important to remember that there is a distinction between the positive integer n and the corresponding element of $\mathbb{F}_p^\times$, which we also denote by n, although a notation like $[n]_p$ would be be more correct. The two objects behave differently under the respective arithmetical operations. For example, the integer 20 is such that $20 \times 20 = 400$, whereas when 20 is taken as an element of $\mathbb{F}_{23}^\times$ (say), we have $20 \times 20 = 9$. The distinction is even more obvious when we consider multiplicative inverses. In order to define the multiplicative inverse of the integer 20 we have to use a different kind of number, the fraction $\frac{1}{20}$. On the other hand, the multiplicative inverse of 20 as an element of $\mathbb{F}_{23}^\times$ is 15, since $20 \times 15 = 1 \pmod{23}$.

From a practical point of view, the significant point is that some, but not all, calculations in $\mathbb{F}_p^\times$ can be done by adapting the familiar methods that we use for ordinary integers. For example, to find 20×20 in $\mathbb{F}_{23}^\times$ we multiply as if 20 was an ordinary integer, and then reduce mod 23 (that is, we find the remainder when 400 is divided by 23):

$$20 \times 20 = 400 = 17 \times 23 + 9 = 9 \pmod{23}.$$

Similarly, in order to find 20^{-1} in $\mathbb{F}_{23}^{\times}$ we can also use ordinary arithmetic, by applying the Euclidean algorithm (Section 13.3). This gives the result $20^{-1} = 15$, which can be checked by another familiar calculation:

$$20 \times 15 = 1 + 13 \times 23.$$

However, there are some calculations in $\mathbb{F}_p^{\times}$ for which (as yet) no one has found a way to employ the familiar algorithms of arithmetic. The most important example arises from the following definition.

Definition 14.4 (Primitive root)

A *primitive root* of a prime p is an element r of the group $\mathbb{F}_p^{\times}$ such that the $p-1$ powers of r

$$r,\ r^2,\ r^3,\ \ldots,\ r^{p-1}$$

are all distinct, and therefore comprise all the elements of $\mathbb{F}_p^{\times}$.

A famous result (conjectured by Euler, and proved by Legendre and Gauss) guarantees that for every prime p there is at least one primitive root r.

Example 14.5

Find all the primitive roots of 11.

Solution Clearly 1 is not a primitive root, so the first candidate is 2. A simple calculation produces the table

2	2^2	2^3	2^4	2^5	2^6	2^7	2^8	2^9	2^{10}
2	4	8	5	10	9	7	3	6	1

Thus 2 is a primitive root. In general, a primitive root r must have the property that r^{10} is the smallest power of r which is equal to 1, and the table enables us to check this condition. For example, 3 is not a primitive root, since

$$3 = 2^8, \quad \text{and so} \quad 3^5 = 2^{40} = (2^{10})^4 = 1^4 = 1.$$

Using similar arguments it is easy to show that 4, 5, 9, and 10 are not primitive roots, whereas 6, 7, and 8 are. Hence there are four primitive roots $2, 6, 7, 8$. This is a particular case of the general result (Exercise 14.6) that the number of primitive roots of p is $\phi(p-1)$.

EXERCISES

14.5. Show that 3 is a primitive root of 17, and hence find all the primitive roots of 17.

14.6. Use your calculations from the previous exercise to find the inverse of every element of $\mathbb{F}_{17}^{\times}$.

14.7. Alice and Bob have agreed to express their messages as sequences of elements of $\mathbb{F}_p^{\times}$. They choose keys a, b $(1 \leq a, b \leq p-1)$, and their encryption functions are

$$E_a(m) = m^a, \qquad E_b(m) = m^b \quad (m \in \mathbb{F}_p^{\times}).$$

What conditions on a and b must be satisfied in order that there should exist suitable decryption functions $D_{a'}$ and $D_{b'}$ of a similar form? If these conditions are satisfied, explain how Alice and Bob can avoid the key distribution problem by using the double-locking procedure described in Section 12.6.

14.8. Let r be a primitive root of p. Prove that r^k is also a primitive root if and only if $\gcd(k, p-1) = 1$. Deduce that the number of primitive roots of p is $\phi(p-1)$.

14.4 The discrete logarithm

Suppose a prime p and a primitive root r are given, and consider the 'exponential' function $x \mapsto r^x$. Given x, we can compute the function value $y = r^x$ by using the repeated-squaring algorithm (Section 13.3). The obvious way to attack the reverse problem,

$$\text{given } y, \text{ find } x \text{ such that } r^x = y,$$

is by 'brute force' – working out the powers of r in turn, until the correct value of x is found. Methods of solving the DLP that are better than brute force are known, but there is as yet no general method that will work for large numbers.

The general problem is known as the *Discrete Logarithm Problem* or *DLP*. The name arises from the fact that (in real analysis) the logarithm function is the inverse of the exponential function. In our case we can define the *discrete* logarithm of $y \in \mathbb{F}_p^{\times}$, with respect to the primitive root r, to be the integer x such that

$$r^x = y \quad \text{in } \mathbb{F}_p^{\times} \qquad \text{and} \quad 1 \leq x \leq p-1.$$

The MAPLE command `>mlog(y,r,p)` attempts to find the logarithm of y to base r, in $\mathbb{F}_p^\times$. However, the current implementation is such that it will usually fail to produce an answer for large values of p. Intuitively, the difficulty arises because the behaviour of the exponential function $x \mapsto r^x$ on $\mathbb{F}_p^\times$ is quite unlike the behaviour of the corresponding function on the field of real numbers. In the real case the function is continuous: given that $2^4 = 16$ (for example), then we can infer that the real number x for which $2^x = 17$ is close to 4. By comparison, the behaviour of $x \mapsto 2^x$ on $\mathbb{F}_p^\times$ is quite irregular, even for small values of p, such as $p = 11$ (Example 14.5).

Example 14.6

Show that 5 is a primitive root of 23, and find the logarithms to base 5 of 2 and 3 (mod 23).

Solution In order to show that 5 is a primitive root it is sufficient to show that the order of 5 (the least value of x such that $5^x = 1$) is 22. It follows from elementary Group Theory that the order must be a divisor of 22, so we have only to check $x = 2$ and $x = 11$. Now

$$5^2 = 2 \qquad \text{and} \quad 5^{11} = (5^2)^5 \times 5 = 2^5 \times 5 = 9 \times 5 = 45 = -1.$$

Since $5^2 \neq 1$ and $5^{11} \neq 1$, the order of 5 is indeed 22.

Since $5^2 = 2$, it follows that $\log_5 2 = 2$. But this does not help us to find $\log_5 3$, so brute force is the simplest way:

$$5^2 = 2,\ 5^3 = 10,\ 5^4 = 4,\ \ldots,\ 5^{16} = 3,\ \ldots .$$

Hence $\log_5 3 = 16$.

As an illustration, here is a simple application of the DLP in computer science, a method of storing passwords on a server.

Suppose a set of users wish to login securely to a server. The administrator chooses a prime p and a primitive root r, and each user chooses a password that can be kept secure, either by memorizing it or recording it in a safe way. The password for each user (Alice) is converted by some automatic process into an element π_A of $\mathbb{F}_p^\times$. It should be noted that if the 'raw' passwords are chosen too predictably – for example, if Alice chooses a real word such as WONDERLAND – then the system will be vulnerable to an attack based on searching a dictionary.

The administrator then calculates $\sigma_A = r^{\pi_A}$ (mod p) and stores the list of pairs (A, σ_A) on the server. The values of p and r will also be kept on the server. Alice can authenticate her identity by sending her password, which is converted to the form π_A, so that the server can check that $r^{\pi_A} = \sigma_A$ (mod p).

This system provides a degree of security, even if a hacker succeeds in obtaining all the information stored on the server, including p, r, and σ_A. Given this information it is still hard to determine Alice's password π_A, because that would entail solving the DLP, $r^x = \sigma_A \pmod{p}$.

EXERCISES

14.9. Verify that 2 is a primitive root of 19, and find the logarithms (to base 2) of 12 and 15 (mod 19).

14.10. An incompetent administrator has set up a password scheme based on $p = 101$ and the primitive root $r = 2$. Eve hacks into the server and discovers these values, and the value stored against Alice's name, which is $\sigma_A = 27$. What is Alice's password π_A?

14.11. Let $\log_r x$ denote the logarithm of x to base r, where r is a primitive root for a given prime p. Prove that

$$\log_r(ab) = \log_r a + \log_r b \pmod{p-1}.$$

14.5 The ElGamal cryptosystem

In 1985 ElGamal showed that a public key cryptosystem can be based on the Discrete Log Problem. It is assumed that all messages are expressed as elements of $\mathbb{F}_p^\times$ in a standard way (for long messages several elements may be needed), and a fixed primitive root r of p is known to all users. Each user, such as Bob, chooses a private key $b' \in \mathbb{N}$, and computes his public key $b \in \mathbb{F}_p^\times$ by the rule $b = r^{b'}$.

An important feature of ElGamal's system is that the encryption of a message involves a *token*, $t \in \mathbb{N}$, which is chosen randomly each time the system is used. When Alice wishes to send Bob a message m she chooses $t \in \mathbb{N}$, and using the publicly available values of r and b, she applies the encryption function

$$E_b(m, t) = (r^t, mb^t).$$

Thus Bob receives ciphertext in two parts: the first part is the 'leader' $\ell = r^t$, and the other part is the encrypted message $c = mb^t$.

Lemma 14.7

The decryption function $D_{b'}$ defined by

$$D_{b'}(\ell, c) = c\,(\ell^{-1})^{b'}.$$

is the left-inverse of the encryption function E_b defined above, for all $t \in \mathbb{N}$.

Proof

We have to check that $D_{b'}(E_b(m,t)) = m$ for all $m \in \mathbb{F}_p^\times$ and all $t \in \mathbb{N}$:

$$\begin{aligned} D_{b'}(E_b(m,t)) = D_{b'}(r^t, mb^t) &= (mb^t)((r^t)^{b'})^{-1} \\ &= (mb^t)((r^{b'})^t)^{-1} \\ &= (mb^t)(b^t)^{-1} \\ &= m. \end{aligned}$$

□

Example 14.8

Suppose $p = 1009, r = 102$, and Bob has chosen the private key $b' = 237$.
(i) What is Bob's public key?
(ii) If Alice wishes to send the message $m = 559$ to Bob, and chooses the token $t = 291$, what ciphertext will Bob receive?
(iii) How will Bob decrypt the ciphertext?

Solution (i) Bob's public key is

$$b = 102^{237} = 854.$$

(ii) If Alice uses E_b to send him the message $m = 559$ with the token $t = 291$, then Bob will receive the ciphertext

$$(r^t, mb^t) = (102^{291}, 559 \times 854^{291}) = (658, 23).$$

(iii) On receiving $(658, 23)$ Bob will apply his decryption function $D_{b'}(\ell, c) = c\,(\ell^{b'})^{-1}$, obtaining

$$23 \times (658^{237})^{-1} = 23 \times 778^{-1} = 23 \times 463 = 559.$$

Clearly, the security of the ElGamal system depends on the difficulty of the DLP. If Eve can solve the DLP to find x such $r^x = b \pmod{p}$ then Bob's private key b' is x, and Eve can apply the decryption function $D_{b'}$. Other forms of attack are also possible, and the parameters must be chosen with these in mind. For example, if Alice always uses the same value of t, then the system is vulnerable (see Exercise 14.14).

EXERCISES

14.12. Suppose I am using an experimental version of the ElGamal cryptosystem, with $p = 10007$ and $r = 101$. I choose my private key to be $k' = 345$. What is my public key?

14.13. Continuing with the set-up described in the previous exercise, suppose another user wishes to send me the message $x = 332$, and chooses the token $t = 678$. What is the encrypted form (ℓ, c) of the message, as I receive it? Verify that my rule for decryption correctly recovers the original message.

14.14. Show that if Alice always uses the same value of t then the ElGamal system can be broken by a known plaintext attack, with one piece of plaintext m_1 and the corresponding ciphertext (ℓ_1, c_1).

14.6 The Diffie-Hellman key distribution system

In Section 12.6 we noted that one of the problems of symmetric key cryptography is the distribution of the keys. If Alice and Bob wish to communicate, they must first agree on a key, and that agreement appears to require a separate form of communication. In 1976 Diffie and Hellman proposed a key-distribution system that avoids this problem.

Suppose a set of users wish to communicate in pairs, using a symmetric key system. Each pair, such as Alice and Bob, must choose a key k_{AB} in such a way that no other user has direct knowledge of it.

The Diffie-Hellman system begins by assuming that a large prime p and a primitive root r have been selected, and they are public knowledge. Alice then chooses a private key $a' \in \mathbb{N}$, and computes

$$a = r^{a'}.$$

Alice declares a to be her public key. Similarly, Bob chooses $b' \in \mathbb{N}$ and declares his public key $b = r^{b'}$.

When Alice and Bob wish to communicate, they use the key

$$k_{AB} = r^{a'b'}.$$

Observe that Alice can easily calculate k_{AB} using her own private key a' and Bob's public key b, since $k_{AB} = b^{a'}$. Similarly, Bob can calculate k_{AB} using the formula $a^{b'}$.

On the other hand, suppose Eve wishes to discover k_{AB}. She knows the values a, b, and so it would be enough to solve either of the Discrete Log Problems

$$r^x = a, \qquad r^x = b,$$

which would give her a' or b'. But if the numbers are large enough, these problems are thought to be hard. (There might also be a quite different method of solving this particular problem, but as yet no one has found it.)

Example 14.9

Suppose Fiona, Georgina, and Henrietta have agreed to encrypt their text messages using the Diffie-Hellman system with $p = 101$ and $r = 2$. They have chosen the private keys $f' = 13$, $g' = 21$, $h' = 30$, respectively.

(i) What common information will be stored in the directory of each girl's phone?
(ii) What key will be used for messages between Georgina and Henrietta, and how do they obtain it?
(iii) How could Fiona eavesdrop on messages between Georgina and Henrietta?

Solution (i) Each directory will contain the values $p = 101$, $r = 2$, and the public keys f, g, h, computed by the rules $f = 2^{13}$, $g = 2^{21}$, $h = 2^{30}$, as follows.

f	g	h
12	89	17

(ii) The key for communication between Georgina and Henrietta is

$$k_{GH} = 2^{21\times 30} = 17.$$

Georgina can calculate this as $h^{g'} = 17^{21}$ and Henrietta as $g^{h'} = 89^{30}$.
(iii) If Fiona wishes to discover k_{GH} she could try to find either g' or h', by solving one of the equations $2^x = 89, 2^x = 17$, or she could use some other (as yet unknown) method.

EXERCISES

14.15. Four people, A, B, C, D, have chosen to communicate using the Diffie-Hellman system, with $p = 149$ and $r = 2$. If A has chosen the private key 33, what is her public key?

14.16. Continuing with the set-up described in the previous exercise, suppose the public keys of B, C, D are $46, 58, 123$ respectively. What key should A use to communicate with B, and how does she obtain it? If A wishes to discover the key used for communication between C and D, what Discrete Log Problems might she try to solve?

14.7 Signature schemes

In this section we return to the aspects of cryptography discussed at the beginning of this chapter, with particular emphasis on the topic of *non-repudiation*.

Suppose Alice wishes to send the message m to Bob, who would like to have irrefutable evidence that the message really does come from Alice. Clearly, she must include in the message her 'real' name, so that Bob will be aware of the purported sender. But she can also add a *signature*, $y = s(m)$, in a form that confirms her identity. Her definition of $s(m)$ must remain private, otherwise anyone could forge her signature. In order to verify that a message m accompanied by a signature y is really signed by Alice, Bob must be able to decide whether or not $s(m) = y$ without knowing precisely how Alice constructed the function s. Note that this procedure is quite different from the traditional use of hand-written signatures. In that case a signature is a fixed object, independent of the message, whereas here the relationship between the signature and the message is crucial. Also, it is useful for Alice to be able to use several different signatures.

The context for the following definition is a communication system in which a typical user A has a public key a and a private key a'.

Definition 14.10 (Signature scheme)

A *signature scheme* for a set $\mathcal{M}$ of messages comprises a set $\mathcal{Y}$ of *signatures* and

- for each user A, a set $\mathcal{S}_A$ of *signature functions* $s : \mathcal{M} \to \mathcal{Y}$, such that A can calculate $s(m)$ for $s \in \mathcal{S}_A$ and $m \in \mathcal{M}$ using the private key a';
- a *verification algorithm* that, given $m \in \mathcal{M}$ and $y \in \mathcal{Y}$, enables any other user to check the truth of the statement

$$s(m) = y \text{ for some } s \in \mathcal{S}_A$$

using only the public key a. We shall say that y is a *valid signature* for a message m from A if the statement is true.

In order to see how the scheme works, we begin by explaining how, under certain conditions, a public-key cryptosystem can be adapted for use as a signature scheme. Suppose a cryptosystem is given and, for each user A, let $\mathcal{S}_A$ consist of a single signature function s, which is actually the user's decryption function $D_{a'}$. Since we require a signature function to have domain $\mathcal{M}$ and range $\mathcal{Y}$, and $D_{a'}$ is in fact a function from $\mathcal{C}$ to $\mathcal{M}$, we must assume that $\mathcal{M} = \mathcal{C} = \mathcal{Y}$ here.

Using $D_{a'}$ as the signature function, Alice can send the signed message

$$(m, D_{a'}(m)) \; = \; (m, y)$$

to Bob in the usual way, using his (public) encryption function E_b to encrypt both parts. So Bob receives the ciphertext in the form

$$(E_b(m), E_b(y)).$$

He decrypts this ciphertext using his own decryption function $D_{b'}$, and recovers (m, y). The first part m is meaningful plaintext, but the second part y is apparently unintelligible. However Bob knows that the message purports to come from Alice, because the meaningful plaintext says so. He also knows Alice's public key a, and his problem is how to check that y is a valid signature, using a, but not her private key a'.

Suppose Bob applies E_a to y: $E_a(y) = E_a(D_{a'}(m))$. If it happens that $E_a(D_{a'}(m)) = m$ then Bob will have another copy of m. Since only Alice could have constructed $D_{a'}(m)$, Bob is sure that Alice really is the sender of m. What is more, he can prove it, by producing both m and $D_{a'}(m)$.

However, the condition that $E_a(D_{a'}(m)) = m$ for all $m \in \mathcal{M}$ does not hold in general. We do know that $D_{a'}$ is the left-inverse of E_a,

$$D_{a'}(E_a(m)) = m \quad \text{for all } m \in \mathcal{M},$$

because that is the basic property of a cryptosystem, but the condition required here involves the operators in reverse order. In other words, the condition is that the operators *commute*. To summarize, we have proved the following theorem.

Theorem 14.11

Suppose there exists a public-key cryptosystem with the properties that $\mathcal{M} = \mathcal{C}$ and, for each user A, the functions E_a and $D_{a'}$ commute. Then there is an associated signature scheme in which $\mathcal{Y} = \mathcal{M}$ and

- $\mathcal{S}_A$ contains just one signature function $D_{a'}$;
- any user can check that y is a valid signature for the message m from A by making the test: is $E_a(y) = m$?

Example 14.12

Show that the commutativity property holds in the RSA cryptosytem.

Solution In RSA the encryption and decryption functions for A are given by $E_{n,e}(x) = x^e$, $D_{n,d}(x) = x^d$, where $x \in \mathbb{Z}_n$. Hence

$$E_{n,e}(D_{n,d}(x)) = (x^d)^e = (x^e)^d = D_{n,d}(E_{n,e}(x)).$$

This result shows that RSA can be used as a signature scheme. In practice, the signature is usually applied to a hash of the message, $h(m)$ (represented as an element of $\mathbb{Z}_n$ in some standard way), rather than to m itself. Alice's signature is

$$y = h(m)^d \pmod{n},$$

and Bob's verification test is

$$y^e = h(m) \pmod{n}?$$

On the other hand, the commutativity property does not hold for the ElGamal cryptosystem, so that system cannot be used without modification. ElGamal himself explained how to make the changes required.

ElGamal's signature scheme is defined in the following theorem. As with the cryptosystem, all users are assumed to know a prime p and a primitive root r, and the set of messages is taken to be $\mathcal{M} = \mathbb{F}_p^\times$. The user A has a private key $a' \in \mathbb{N}$ and a public key $a \in \mathbb{F}_p^\times$, related by $a = r^{a'}$. There are many signature functions for A, rather than just one. Specifically, the set $\mathcal{S}_A$ contains functions s_t that depend upon a suitably chosen parameter $t \in \mathbb{N}$:

$$\mathcal{S}_A = \{s_t \mid 1 \leq t \leq p-1 \text{ and } \gcd(t, p-1) = 1\}.$$

In the following theorem it is useful to distinguish between an element of $\mathbb{F}_p^\times$ and the integer that represents it. We adopt the temporary notation that $|x|$ denotes the unique integer that represents $x \in \mathbb{F}_p^\times$ and satisfies $1 \leq |x| \leq p-1$.

Theorem 14.13 (ElGamal signature scheme)

Let t and u be integers such that $1 \leq t, u \leq p-1$ and

$$tu = 1 \pmod{p-1}.$$

Let $\mathcal{Y} = \mathbb{F}_p^\times \times \mathbb{N}$ and define the signature function $s_t : \mathcal{M} \to \mathcal{Y}$ by

$$s_t(m) = (i, j) \quad \text{where } i = r^t,\ j = u(|m| - a'|i|).$$

Then (i, j) is a valid signature for the message m from A if and only if

$$a^{|i|} i^j = r^{|m|} \quad \text{in } \mathbb{F}_p^\times.$$

Proof

If (i, j) is a valid signature, then $i = r^t$ for some value of t. Thus

$$a^{|i|} i^j \;=\; r^{a'|i|} r^{tj} \;=\; r^{a'|i|+tj}.$$

Since $tu = 1 \pmod{p-1}$ we have

$$a'|i| + tj \;=\; a'|i| + tu(|m| - a'|i|) \;=\; |m| \pmod{p-1},$$

as claimed.

Conversely, suppose m and (i, j) are such that $a^{|i|} i^j = r^{|m|}$. Since i is in $\mathbb{F}_p^\times$ we have $i = r^t$ for some t such that $1 \leq t \leq p-1$. Hence

$$r^{|m|} = a^{|i|} i^j = r^{a'|i|} r^{tj} = r^{a'|i|+tj}.$$

Since r has order $p-1$ this implies that

$$|m| = a'|i| + tj \pmod{p-1}, \quad \text{that is} \quad j = u(|m| - a'|i|).$$

□

EXERCISES

14.17. Suppose the ElGamal signature scheme is used with $p = 23$ and $r = 5$. Alice's public key is $a = 6$. Is $(2, 4)$ a valid signature for the message $m = 6$ from Alice?

14.18. Show that the number of valid signatures for a given message in the ElGamal scheme is $\phi(p-1)$. In the previous exercise, what are the possible values of i in a valid signature (i, j), and why is $i = 2$ not one of them?

14.19. Suppose that, in the ElGamal system, Eve knows Alice's public key a but not her private key a', and she has a fake message m that she wants Bob to think has been sent by Alice. There are two simple strategies that might allow her to construct a valid signature (i, j).

(1) Choose i randomly and attempt to calculate the correct j.

(2) Choose j randomly and attempt to calculate the correct i.

Show that the first strategy is equivalent to solving a DLP. What equation must be solved in order to succeed with the second strategy? [It is not known how this problem is related to the DLP.]

14.20. Alice and Bob have agreed to use the RSA signature scheme (as described above), and Alice's public key is $n = 187, e = 3$. She has arranged with Bob that they will use a hash function h with values in ${\mathbb{F}_2}^8$, regarded as the binary representation of numbers mod 187. (For example, 00101100 is regarded as 44.) Bob has received two messages purporting to come from Alice. For one the hash is 00001011, and the signature is 9, and for the other the hash is 00010101 and the signature is 98. Write down Bob's verification test and decide whether either of the messages could be genuine.

Further reading for Chapter 14

The relationship between cryptography and complexity theory is explained thoroughly in the book by Talbot and Welsh [**14.3**]. The surveys of cryptography listed at the end of Chapter 12, especially those by Menezes *et al.* [**12.3**], Stinson [**12.5**], and Trappe and Washington [**12.6**], also contain a great deal of relevant material.

The Diffie-Hellman key-distribution system based on the DLP was described in the original paper on public key cryptography [**13.4**]. Soon afterwards Pohlig and Hellman showed that solving the DLP on any cyclic group of order n could be reduced to solving the DLP on the cyclic groups of order f, where f is a prime factor of n. This implies that when applications that rely on the DLP on $\mathbb{F}_p^\times$ are designed, the prime p should be chosen so that $p - 1$ has very few prime factors. Designers must also be aware of the attack by *index calculus.* This is a method specific to the group $\mathbb{F}_p^\times$ that can produce results if p is chosen unwisely; it is the described in the cryptography texts mentioned above.

ElGamal's algebraic formulae for $\mathbb{F}_p^\times$ were published in 1985 [**14.1**]. In the next chapter we shall explain how the same formulae can be applied more generally.

14.1 T. ElGamal. A public key cryptosystem and a signature scheme based on discrete logarithms. *IEEE Trans. on Information Theory* 31 (1985) 469-472.

14.2 S.C. Pohlig and M.E. Hellman. An improved algorithm for computing logarithms over $GF(p)$ and its cryptographic significance. *IEEE Trans. on Information Theory* 24 (1978) 106-110.

14.3 J. Talbot and D. Welsh. *Complexity and Cryptography.* Cambridge University Press (2006).

15
Elliptic curve cryptography

15.1 Calculations in finite groups

In this chapter we continue to adopt the *cryptographic perspective* developed in the earlier chapters. This means that we stress the link between mathematical theory and practical calculation which, as we have seen, is fundamental in modern cryptography. The specific system that we consider is based on groups that arise in the theory of elliptic curves, a topic that fascinated mathematicians for over a century before it first found practical application in the 1980s.

From the cryptographic perspective, it is worth stressing that a group consists of two things: a set of elements G, and an operation $*$ defined on pairs of elements (h, k) in G. When we say that a group $(G, *)$ is 'given', we mean that we know how to represent h and k in a definite way, and how to calculate $h * k$ using this representation. Ultimately, it must be possible to reduce these calculations to operations on strings of bits, although it is usually convenient to use a more familiar notation, such as the standard decimal notation for integers.

For example, suppose we are working in $\mathbb{F}_p^\times$, the group of nonzero elements of $\mathbb{F}_p$, under multiplication. Then the elements of $\mathbb{F}_p^\times$ are represented by natural numbers $1, 2, \ldots, p-1$ in decimal notation, and $h \times k$ can be calculated by applying the familiar 'long multiplication' algorithm, followed by 'long division' to find the remainder when the result is divided by p.

N. L. Biggs, *An Introduction to Information Communication and Cryptography*,
DOI: 10.1007/978-1-84800-273-9_15,

Definition 15.1 (Cyclic group, generator)

The group $(G, *)$ is a *cyclic group of order* n, *with generator* g if $|G| = n$ and there is an element $g \in G$ such that the elements of the group are equal to

$$g, g^2, g^3, \ldots, g^n,$$

in some order. (It follows that the group is abelian, and g^n is the identity element.)

The powers of any element $h \in G$ are defined recursively by the rule

$$h^1 = h, \qquad h^i = h * h^{i-1} \; (i \geq 2).$$

Although the definition suggests that $i - 1$ applications of the $*$ operation are needed in order to calculate h^i, the repeated squaring algorithm (Section 13.3) allows h^i to be calculated much more efficiently. Furthermore, if G has n elements, every $h \in G$ is such that h^n is equal to the identity element. Thus calculating the inverse of h can, if necessary, be done by using the rule $h^{-1} = h^{n-1}$.

The fact that $\mathbb{F}_p^\times$ is a cyclic group of order $p - 1$ is a consequence of the famous result that there is a primitive root r (a generator of $\mathbb{F}_p^\times$) for every prime p. However, there are two problems. Finding a primitive root r for a given p is not trivial, and the correspondence between the powers of r and the elements of $\mathbb{F}_p^\times$ is complicated. The latter problem is just the Discrete Logarithm Problem discussed in the previous chapter. As we shall see, both these problems occur more generally in elliptic curve cryptography.

The following example illustrates the fact that, without some additional information, finding a generator for a cyclic group may require 'brute force' methods.

Example 15.2

Let $*$ denote multiplication of 2×2 matrices. Find g such that the following set of six matrices forms a cyclic group $(G, *)$ with generator g.

$$\begin{pmatrix} 1 & 0 \\ 0 & 1 \end{pmatrix} \quad \begin{pmatrix} -1 & 0 \\ 0 & -1 \end{pmatrix} \quad \begin{pmatrix} \frac{1}{2} & \frac{\sqrt{3}}{2} \\ -\frac{\sqrt{3}}{2} & \frac{1}{2} \end{pmatrix}$$

$$\begin{pmatrix} -\frac{1}{2} & \frac{\sqrt{3}}{2} \\ -\frac{\sqrt{3}}{2} & -\frac{1}{2} \end{pmatrix} \quad \begin{pmatrix} -\frac{1}{2} & -\frac{\sqrt{3}}{2} \\ \frac{\sqrt{3}}{2} & -\frac{1}{2} \end{pmatrix} \quad \begin{pmatrix} \frac{1}{2} & -\frac{\sqrt{3}}{2} \\ \frac{\sqrt{3}}{2} & \frac{1}{2} \end{pmatrix}$$

Solution If we are unaware of the geometrical significance of the matrices, we must proceed by working out the orders of the matrices. The first two matrices have orders 1 and 2 respectively, and clearly they are not generators. However, if we take the third matrix to be g, then it turns out that the six given matrices are equal to $g, g^2, g^3, g^4, g^5, g^6$ (but obviously not in the given order). Thus g is a generator.

Suppose the problem of finding a generator g in a cyclic group G has been solved. Then we might hope to use the correspondence between the elements of G and the powers of g to simplify calculations in the group. Thus in order to calculate $h * k$ we can write $h = g^i$, and $k = g^j$, calculate $i + j$ using ordinary addition, and set $h * k = g^{i+j}$. But this method assumes that we can find i and j, the 'logarithms' of h and k. That is precisely the Discrete Log Problem (DLP) for $(G, *)$:

given a generator $g \in G$ and any $h \in G$, find $i \in \mathbb{N}$ such that $g^i = h$.

In some cases this problem may be easy (Exercise 15.3), but in general it is hard.

EXERCISES

15.1. Show that the operation of multiplying complex numbers makes the following set of numbers into a cyclic group, and find a generator.

$$1,\ -1,\ i,\ -i,\ \frac{1}{\sqrt{2}}(1+i),\ \frac{1}{\sqrt{2}}(1-i),\ \frac{1}{\sqrt{2}}(-1+i),\ \frac{1}{\sqrt{2}}(-1-i).$$

15.2. In Example 14.5 we constructed a table of logarithms to base 2 in the cyclic group $\mathbb{F}_{11}^{\times}$. Explain how the table can be used to show that $5 \times 10 = 6$ in this group.

15.3. Show that, for any positive integer n, the group $(\mathbb{Z}_n, +)$ (the integers mod n under addition) is cyclic, and find a generator. Explain why the DLP is trivial in this case, however large n may be.

15.2 The general ElGamal cryptosystem

The ElGamal systems described in Chapter 14 can be extended to any given cyclic group $(G, *)$. Users of the general system are assumed to know that G is cyclic, and that a certain specified element $g \in G$ is a generator. It is also assumed that they express plaintext and ciphertext messages as elements of G in some standard way.

In the general ElGamal cryptosystem, each user, such as Bob, chooses a private key $b' \in \mathbb{N}$, and computes his public key $b \in G$ by the rule $b = g^{b'}$. When Alice wishes to send Bob a message $m \in G$ she chooses a token $t \in \mathbb{N}$, and applies the encryption function

$$E_b(m, t) = (g^t, m * b^t).$$

Bob receives ciphertext in two parts: the first part is the 'leader' $\ell = g^t$, and the other part is the encrypted message $c = m * b^t$. Bob's decryption function is

$$D_{b'}(\ell, c) = c * (\ell^{-1})^{b'}.$$

Using essentially the same algebra as in Lemma 14.7, it is easy to check that $D_{b'}(E_b(m, t)) = m$ for all $m \in G$, and all $t \in \mathbb{N}$:

$$\begin{aligned} D_{b'}(E_b(m,t)) = D_{b'}(g^t, m * b^t) &= (m * b^t) * ((g^t)^{-1})^{b'} \\ &= m * b^t * ((g^{b'})^{-1})^t \\ &= m * b^t * (b^{-1})^t \\ &= m. \end{aligned}$$

EXERCISES

15.4. Let $(G, *)$ be a cyclic group of order 43, with generator g, and suppose Bob's private key is 10. What is Bob's public key, and what is his decryption function? If Alice wishes to send the message $m \in G$ to Bob, and chooses $t = 7$, what ciphertext does Bob receive? Check that his decryption function correctly recovers m. (All working should be expressed in terms of the 'variables' g and m.)

15.5. Alice and Bob are experimenting with an ElGamal system based on the multiplicative group $G = \mathbb{F}_{17}^{\times}$, with generator $g = 3$. Bob's public key is 13. Alice wishes to send the message $m \in \mathbb{F}_{17}^{\times}$ to Bob. What encryption function should she use? Find Bob's private key, write down his decryption function, and verify that it correctly recovers the message m.

15.6. In Exercise 15.3 we noted that $(\mathbb{Z}_n, +)$ is cyclic group with generator 1. Show that in the corresponding ElGamal system $b' = b$, and verify that the decryption function $D_{b'}$ is the left-inverse of E_b.

15.3 Elliptic curves

In the rest of the book we shall consider fields F with the property that $2x = 0$ only if $x = 0$ (so that $F \neq \mathbb{F}_2$, for example). This condition is expressed by the statement that F does not have *characteristic* 2. The reason for excluding fields with characteristic 2 is that the algebra takes a slightly different form in that case.

Definition 15.3 (Elliptic curve)

Let F be a field which does not have characteristic 2. An *elliptic curve* over F is a set of 'points' $(x, y) \in F^2$ that satisfy an equation of the form

$$y^2 = x^3 + \alpha x + \beta \qquad (\alpha, \beta \in F),$$

together with one additional 'point', which is denoted by I and called the *point at infinity*.

For cryptographic purposes we shall require that the field F is finite, but the same constructions can be used over any field F which does not have characteristic 2, and it is helpful to begin by looking at an example with $F = \mathbb{R}$, the field of real numbers. In this case the 'points' that form the curve belong to the Euclidean plane $\mathbb{R}^2$, and we can sketch the curve in the usual way. The resulting geometrical picture is very useful.

Example 15.4

Sketch the curve $y^2 = x^3 - x$ over $\mathbb{R}$.

Solution The standard method of curve-sketching is to find points on the curve by choosing $x \in \mathbb{R}$ and calculating the value(s) of y such that $y^2 = x^3 - x$.

When $x < -1$ and when $0 < x < 1$ the expression $x^3 - x$ is negative, and there are no corresponding values of y, since $y^2 \geq 0$.

When $x = -1, 0, 1$, $x^3 - x = 0$ and $y = 0$ is the only possibility. Hence the points $(-1, 0)$, $(0, 0)$, and $(1, 0)$ belong to the curve.

Finally, for each remaining value of x (that is $-1 < x < 0$ and $x > 1$) the expression $x^3 - x$ is positive and there are two corresponding values of y. This means that the curve is symmetrical with respect to the x-axis: if (x, y) is on the curve, then $(x, -y)$ is also on the curve. A sketch is shown in Figure 15.1.

It must be remembered that I, the 'point at infinity', must also be considered. Geometrically, it is helpful to think of I as a point where all vertical lines meet; that is, an infinitely remote point in the vertical direction.

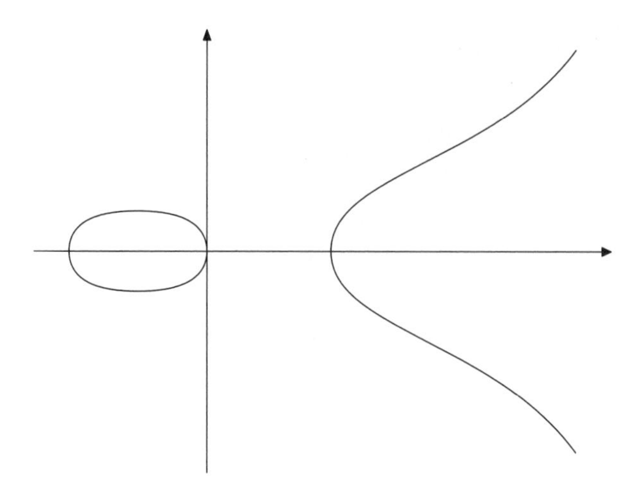

Figure 15.1 A sketch of the curve $y^2 = x^3 - x$ in $\mathbb{R}^2$

Similar calculations can be used when the field is finite, but of course there is no sensible way of 'sketching' the curve.

Example 15.5

Find all the points on the curve $y^2 = x^3 + x$ over $\mathbb{F}_{17}$.

Solution As in the previous example, we consider each value of x and solve the resulting equation for y. For example, when $x = 1$ we require $y^2 = 2$, and in $\mathbb{F}_{17}$ this has two solutions, $y = \pm 6$. (Note that $-6 = 11$ here.)

For each $x \in \mathbb{F}_{17}$ there are $0, 1$, or 2 possible values of y:

$x =$	0	1	2	3	4	5	6	7	8	9
$y =$	0	± 6	–	± 9	0	–	± 1	–	–	–

$x =$	10	11	12	13	14	15	16
$y =$	–	± 4	–	0	± 2	–	± 7

The calculation gives 15 points which, together with I, the point at infinity, comprise the curve.

We shall now explain how an operation $*$ can be defined so that the points of an elliptic curve over a field F form an abelian group. The definition is based on a geometrical construction and is easy to visualize in the case when $F = \mathbb{R}$. But the rules of algebra are the same in any field, so we can translate the construction into familiar coordinate geometry and apply it quite generally.

We begin by specifying that the point at infinity I is the *identity* element, so $P * I = I * P = P$ for all points P. The *inverse* of a point $P = (x, y)$ is $P^{-1} = (x, -y)$: geometrically speaking, P^{-1} is the reflection of P in the x-axis. Note that if P is on the curve, so is P^{-1}. Also, if Q is a point of the form $(x, 0)$ then $Q^{-1} = Q$, so $Q * Q = I$.

The crucial part of the construction is the definition of $P_1 * P_2$, which depends on the fact that the right-hand side of the equation is a polynomial of degree 3. It follows that a straight line $y = \lambda x + \mu$ will generally meet the curve in three points. We define

$P_1 * P_2 = S$ *if and only if the points* P_1, P_2 *and* S^{-1} *are collinear.*

Figure 15.2 shows three collinear points P_1, P_2, S^{-1} and the point $S = P_1 * P_2$.

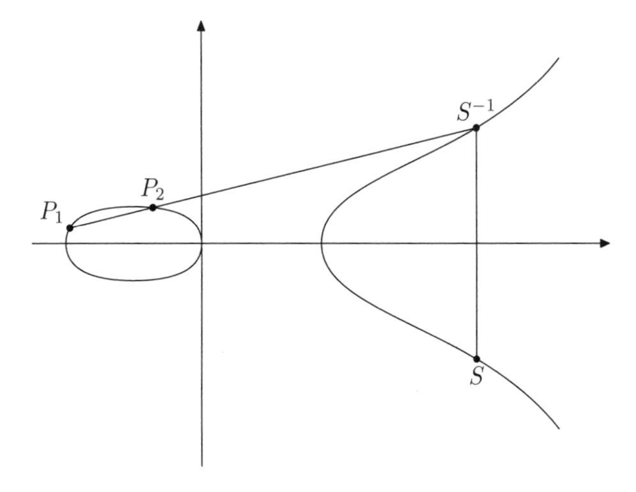

Figure 15.2 Illustrating the group operation

We now translate this construction into coordinate geometry. That is, we find equations for the coordinates of $S = P_1 * P_2$, given the coordinates of P_1 and P_2.

Theorem 15.6

Suppose the points $P_1 = (x_1, y_1)$ and $P_2 = (x_2, y_2)$ belong to an elliptic curve $y^2 = x^3 + \alpha x + \beta$ over a field F which does not have characteristic 2. Then the point $S = P_1 * P_2$ is determined by the following rules:

(i) if $x_1 = x_2$ and $y_1 = -y_2$ then $S = I$;

(ii) in all other cases the coordinates (x_S, y_S) of the point S are given by

$$x_S = \lambda^2 - x_1 - x_2 \qquad y_S = \lambda(x_1 - x_S) - y_1,$$

where λ is defined by

$$\lambda = (y_2 - y_1)(x_2 - x_1)^{-1} \qquad \text{if } x_1 \neq x_2;$$

$$\lambda = (3x_1^2 + \alpha)(2y_1)^{-1} \qquad \text{if } x_1 = x_2,\ y_1 = y_2.$$

Proof

(i) This is the situation when $P_2 = P_1^{-1}$, as defined above.
(ii) Suppose that $x_1 \neq x_2$ and the line through P_1 and P_2 is given by the equation $y = \lambda x + \mu$. Then

$$y_1 = \lambda x_1 + \mu \qquad \text{and} \qquad y_2 = \lambda x_2 + \mu,$$

so that

$$\lambda = (y_2 - y_1)(x_2 - x_1)^{-1}, \qquad \mu = \lambda x_1 - y_1.$$

This line meets the curve $y^2 = x^3 + \alpha x + \beta$ at the points where x satisfies

$$(\lambda x + \mu)^2 = x^3 + \alpha x + \beta, \quad \text{or} \quad x^3 - \lambda^2 x^2 + (\alpha - 2\lambda\mu)x + (\beta - \mu^2) = 0.$$

We know that two of the roots of this equation are x_1 and x_2, and since the sum of the roots is λ^2, there is a third root x_S given by

$$x_S = \lambda^2 - x_1 - x_2.$$

Let $y_S = -(\lambda x_S + \mu)$ so that the point $S^{-1} = (x_S, -y_S)$ is on the line $y = \lambda x + \mu$, and $S = (x_S, y_S)$ is the required point $P_1 * P_2$. Eliminating μ gives

$$y_S = \lambda(x_1 - x_S) - y_1.$$

If $x_1 = x_2$ the definition of λ will not work, because $x_2 - x_1 = 0$ has no inverse. In fact if $x_1 = x_2$ then $y_1^2 = y_2^2$, so there are two possibilities, $y_1 = y_2$ or $y_1 = -y_2$. The second possibility has already been dealt with (case (i)).

If $x_1 = x_2$ and $y_1 = y_2$ then $P_1 = P_2$. In this case, we must determine the line $y = \lambda x + \mu$ that meets the curve in two coincident points. In coordinate geometry over $\mathbb{R}$ we call this a *tangent* to the curve, and determine its *slope* λ by calculus. Because the curve has an algebraic equation, the same results hold in any field F, and the relevant value of λ is as given in the statement of the theorem. The rest of the algebra is as before, with this new value of λ. □

Example 15.7

In Example 15.5 we found that the points $P_1 = (1,6)$ and $P_2 = (11,4)$ belong to the curve $y^2 = x^3 + x$ over $\mathbb{F}_{17}$. Calculate the coordinates of $S = P_1 * P_2$ and $T = P_1 * P_1$.

Solution Taking $P_1 = (1,6)$ and $P_2 = (11,4)$, the coordinates of $S = P_1 * P_2$ are given by

$$\lambda = (4-6) \times (11-1)^{-1} = (-2) \times (10)^{-1} = -2 \times 12 = -24 = 10,$$

$$x_S = 10^2 - 1 - 11 = 88 = 3, \qquad y_S = 10(1-3) - 6 = -26 = 8.$$

Thus $S = (3,8)$. To find $T = P_1 * P_1$ we use the alternative form of λ:

$$\lambda = (3 \times 1^2 + 1) \times (2 \times 6)^{-1} = 4 \times 10 = 40 = 6,$$

$$x_T = 6^2 - 1 - 1 = 34 = 0, \qquad y_T = 6(1-0) - 6 = 0.$$

Thus $T = (0,0)$.

EXERCISES

15.7. Consider the curve $y^2 = x^3 + x$ over $\mathbb{F}_{17}$ discussed in Examples 15.5 and 15.7. Show that $(1,6)$ generates a subgroup of order 4.

15.8. Find explicitly all the points on the elliptic curve $y^2 = x^3 + x$ over $\mathbb{F}_{13}$. (There are 20 of them.) Calculate the coordinates of the points

$$(3,2) * (5,0), \qquad (2,6) * (2,6).$$

15.9. Show that a point $P = (x,y)$ on an elliptic curve has order 2 (that is, $P * P = I$) if and only if $y = 0$.

15.10. Taking $F = \mathbb{R}$, derive the formula for λ given in Theorem 15.6 for the case $P_1 = P_2$.

15.4 The group of an elliptic curve

The $*$ operation endows an elliptic curve with the structure of an abelian group. The relevant properties are almost self-evident, with the exception of the *associative* law: $(A * B) * C = A * (B * C)$. Since we have obtained explicit formulae for the group operation, the associative law can, if necessary, be checked by some rather tedious algebra. (The reader who wants to 'really' understand why it is true is advised to study a more theoretical account of elliptic curves.)

Our aim here is to explain how the group of an elliptic curve can be used in practice, specifically as the basis for an ElGamal cryptosystem. In order to do this, we need to identify a suitable cyclic group, which may be a proper subgroup of the full group, and a generator for it. A very simple example follows.

Example 15.8

For the curve $y^2 = x^3 + 2x + 4$ over $\mathbb{F}_5$, find a cyclic subgroup and a generator of it.

Solution We can tabulate the points on the curve in the usual way:

$x =$	0	1	2	3	4
$y =$	± 2	–	± 1	–	± 1

Thus, remembering I, there are seven points. Since 7 is a prime number, the full group must be cyclic, and any element except I is a generator.

More generally, it would be very useful if we could determine the size and structure of the group of any elliptic curve over a finite field. Sadly, this requires a substantial amount of theory and some nontrivial calculations. But mathematicians have succeed in finding many examples that are suitable for use in practice, and the basic principles are easy to understand.

Lemma 15.9

Let G_E be the group of points on an elliptic curve $E:\ y^2 = x^3 + \alpha x + \beta$ over a finite field F. Define

$m_1 =$ the number of roots of $x^3 + \alpha x + \beta = 0$ in F;

$m_2 =$ the number of $x \in F$ for which $x^3 + \alpha x + \beta$ is a non-zero square in F.

Then

$$|G_E| = m_1 + 2m_2 + 1.$$

Proof

For each $x \in F$ we count how many points (x, y) belong to E. Since F is a field, the equation $y^2 = x^3 + \alpha x + \beta$ has at most two solutions. If $y = \theta$ is a solution then $y = -\theta$ is also a solution, so the number of distinct solutions is 0 or 2 unless $\theta = 0$, when there is just one solution.

In other words there are two solutions when $x^3+\alpha x+\beta$ is a non-zero square in F, one solution when $x^3+\alpha x+\beta=0$ in F, and no solutions when $x^3+\alpha x+\beta$ is not a square in F. Adding 1 for the point at infinity, we have the result. □

The lemma shows that when $F=\mathbb{F}_p$, the largest possible value of $|G_E|$ is $2p+1$, which would occur if $m_2=p$. In fact, one would expect that only about half the values of $x\in\mathbb{F}_p$ are such that $x^3+\alpha x+\beta$ is a square, so that $|G_E|$ will be approximately equal to p. Using this idea, Hasse proved in 1933 (long before elliptic curves became part of cryptography) that

$$p+1-2\sqrt{p} \;\le\; |G_E| \;\le\; p+1+2\sqrt{p}.$$

In the most favourable situation, $|G_E|$ itself is a prime. Then the entire group is a cyclic group, and can be used as framework for systems based on the ElGamal formulae. For example, this happens when E is the curve

$$y^2=x^3+10x+\beta \quad \text{over } \mathbb{F}_p,$$

where

$$\begin{aligned} p &= 2^{160}+7 \\ &= 1461501637330902918203684832716283019655932542983 \\ \\ \beta &= 1343632762150092499701637438970764818528075565078. \end{aligned}$$

It has been shown that

$$\begin{aligned} |G_E| &= 1461501637330902918203683518218126812711137002561 \\ &= p-1314498156206944779554042 2, \end{aligned}$$

and it is easy to check (with MAPLE) that $|G_E|$ is a prime number. It follows that G_E is a cyclic group, and any non-identity element is a generator. So if we follow the prescription described above, we have the basis for cryptosystem. (Note that $|G_E|$ differs from p by a number with 26 digits, whereas p has 53 digits, in accordance with Hasse's theorem.)

Although this favourable situation cannot be expected to occur very often, in practice it is just as useful to be able to find a large prime dividing $|G_E|$. In that case we have a large cyclic subgroup of G_E.

EXERCISES

15.11. Verify explicitly that $(2,1)$ is a generator for the group obtained in Example 15.8.

15.12. Consider elliptic curves of the form $y^2 = x^3 + x + \beta$ over $\mathbb{F}_{11}$. Find three values of β for which m_1 (Lemma 15.9) is 0, 1, 3, respectively.

15.13. Taking $\beta = 6$ in the previous exercise, show that the group of the curve is cyclic, and find a generator for it.

15.14. Let p be an odd prime. Show that the group of the elliptic curve $y^2 = x^3 + x$ over the field $\mathbb{F}_p$ has even order. Find the number m_1 for this group, distinguishing the cases $p = 4s + 1$ and $p = 4s + 3$.

15.15. Any group of order 20 has a cyclic subgroup of order 5. [You are not asked to prove this, but if you are familiar with elementary group theory you may wish to do so.] Determine this subgroup explicitly for the curve described in Exercise 15.8.

15.5 Improving the efficiency of exponentiation

From the cryptographic perspective, it remains to consider the problems that arise when we try to implement a cryptosystem based on an elliptic curve.

The most costly operations in the ElGamal scheme are the *exponentiations* – calculating the powers such as g^t and $(\ell^{-1})^{b'}$ that occur in the encryption and decryption functions. In order to ensure confidentiality the exponents must be numbers with many digits, and the exponentiations, although feasible by the repeated squaring algorithm, are by no means trivial. In fact, that is the main reason why the ElGamal system is commonly used only to distribute keys for a symmetric key system such as AES, in which the calculations are less costly. (Similar remarks apply to RSA, where exponentiation is also a major part of the system.)

The problem of finding good algorithms for exponentiation is therefore significant. It is easy to see that the repeated squaring algorithm is not optimal: for example, it finds g^{15} by calculating

$$g^2 = g * g, \quad g^4 = g^2 * g^2, \quad g^8 = g^4 * g^4,$$

$$g^{12} = g^4 * g^8, \quad g^{14} = g^2 * g^{12}, \quad g^{15} = g * g^{14}.$$

This procedure involves the sequence of powers $1, 2, 4, 8, 12, 14, 15$, which has the property that each term in the sequence except the first is the sum of two (possibly the same) terms that come before it. Any sequence with this property that ends in 15 will produce the the required result.

Definition 15.10 (Addition chain)

An *addition chain of length* r for the positive integer n is a sequence of positive integers $x_0 = 1,\ x_1,\ x_2,\ \ldots,\ x_r = n$ such that, for $i = 1, 2, \ldots, r$ there exist x_j and x_k such that

$$x_i = x_j + x_k \quad (0 \leq j \leq k < i).$$

Clearly, shorter addition chains for n lead to better methods for calculating g^n. For example, the repeated squaring method for g^{15} corresponds to the addition chain of length 6 given above, but there is a shorter addition chain, with length 5: $1, 2, 3, 6, 12, 15$. This corresponds to the multiplications

$$g^2 = g * g, \quad g^3 = g * g^2, \quad g^6 = g^3 * g^3,$$
$$g^{12} = g^6 * g^6, \quad g^{15} = g^3 * g^{12}.$$

A further improvement can be made when inversion is a trivial operation. As we shall explain in the next section, this holds true when G is the group of an elliptic curve. In such cases division (multiplication by a power of g^{-1}) is no more costly than multiplication by g, which motivates the following definition.

Definition 15.11 (Addition-subtraction chain)

An *addition-subtraction chain of length* r for the positive integer n is a sequence of positive integers $x_0 = 1,\ x_1,\ x_2,\ \ldots,\ x_r = n$ such that, for $i = 1, 2, \ldots, r$ there are x_j and x_k such that

$$x_i \ = \ \pm x_j \pm x_k \quad (0 \leq j \leq k < i).$$

In other words, each term in the sequence except the first is the sum or difference of two terms that come before it.

The technique of exponentiation using an addition-subtraction chain is best illustrated by an example.

Example 15.12

Find the optimum method of calculating g^{31}.

Solution The repeated squaring algorithm uses the addition chain 1,2,4,8,16, 24,28,30,31, which has length 8. It is fairly easy to spot an addition chain of length 7: $1, 2, 3, 5, 10, 11, 21, 31$, and some rather tedious analysis will confirm that it is the shortest possible.

However, if subtractions are allowed there is an obvious chain of length 6: $1, 2, 4, 8, 16, 32, 31$, and this is optimal. The corresponding method of calculating g^{31} is

$$g^2 = g * g, \quad g^4 = g^2 * g^2, \quad g^8 = g^4 * g^4,$$
$$g^{16} = g^8 * g^8, \quad g^{32} = g^{16} * g^{16}, \quad g^{31} = g^{-1} * g^{32}.$$

A useful technique for finding a good addition-subtraction chain is based on the *non-adjacent form* of an integer (Exercise 15.18). It leads to an algorithm for exponentiation that is about 10% better than repeated squaring, on average.

EXERCISES

15.16. Write down the addition chain for 127 used in the repeated squaring algorithm. This is a chain with length 12. Show that it is not optimal by finding an addition chain with length 10 for 127.

15.17. Show that the computation of g^{127} can be shortened further if addition-subtraction chains are allowed.

15.18. Consider representations of an integer in the form

$$\sum c_i 2^i \qquad c_i \in \{-1, 0, 1\}.$$

Such a representation is said to be a *non-adjacent form* or *NAF* if $c_i c_{i+1} = 0$ for all $i \geq 0$. Find a NAF for 55 and explain how it can be used to calculate g^{55}.

15.19. Show that every integer has a NAF. [Hint: start from the standard binary representation.] Show also that the NAF is unique.

15.20. Why are addition-subtraction chains not useful for the calculation of g^n when g is an element of $\mathbb{F}_p^\times$?

15.6 A final word

Elliptic curve cryptography is a rapidly growing area of research, and it is possible that future developments will change the picture quite dramatically. Here is a summary of the current state of the art.

- The group of an elliptic curve can be used as the basis for a cryptosystem of the ElGamal type. By making suitable adjustments, elliptic curves can also be used in many other areas of cryptography.
- The ElGamal functions can be calculated fairly efficiently, but the cost of exponentiation (in particular) imposes some constraints in practice.
- It is possible to break an elliptic curve system if there is a method of solving the DLP on the group of the curve but, in general, no such method is known. Other forms of attack may be possible.

We conclude with an explicit example of how the ElGamal formulae can be applied to the group of an elliptic curve. First, we must decide how to represent the elements of the group. For a curve E defined over a prime field $\mathbb{F}_p$, an element of G_E is a pair (x,y) with $x,y \in \mathbb{F}_p$. So we can regard x and y as integers in the range $0 \le x,y \le p-1$. Furthermore, when the right-hand side of the equation is a non-zero square there are exactly two values of y that satisfy the equation, and they can be written uniquely as $\pm\theta$, where θ satisfies $1 \le \theta \le \frac{1}{2}(p-1)$. Since θ is determined by E, in order to store (x,y) it is only necessary to store x as an element of $\mathbb{F}_p$, together with a single bit that determines whether the relevant sign is $+$ or $-$. Incidentally, this observation justifies the use of addition-subtraction chains for exponentiation, since inversion in G_E is a trivially easy operation, the inverse of (x,y) being $(x,-y)$.

Let E denote the curve

$$y^2 = x^3 + x + 4 \quad (x,y \in \mathbb{F}_{23}).$$

We shall use the notation $x+$ and $x-$ for the points (x,y) with $y = \pm\theta$, $1 \le \theta \le 11$: for example, $0+$ stands for $(0,2)$ and $0-$ stands for $(0,-2)$. Substituting $x = 0,1,2,\ldots,22$ in turn, we find that x^3+x+4 is never zero, so that $m_1 = 0$, and it is a square when

$$x = 0, 1, 4, 7, 8, 9, 10, 11, 13, 14, 15, 17, 18, 22,$$

so that $m_2 = 14$. Hence the order of the group G_E is $2 \times 14 + 1 = 29$. Since this number is prime, the group is cyclic, and any element except I can be taken as the generator g.

Conveniently, the group G_E has the right number of symbols to represent the `english` alphabet, extended to include the comma and full stop. Although this number is far too small to provide security in serious applications, it can be used to send messages that are unintelligible to the vast majority of people. To do this, we need to establish a 'standard' correspondence between the 29 elements of the group and the symbols of the extended `english` alphabet. Here is a suitable correspondence.

I	0+	0−	1+	1−	4+	4−	7+	7−	8+
⊔	A	B	C	D	E	F	G	H	I

8−	9+	9−	10+	10−	11+	11−	13+	13−	14+
J	K	L	M	N	O	P	Q	R	S

14−	15+	15−	17+	17−	18+	18−	22+	22−
T	U	V	W	X	Y	Z	,	.

Suppose I am using G_E with generator $g = 0+$, and my public key is $b = 7-$. If you have read this book carefully, you will be able to construct a table of powers of g.

i	2	3	4	5	6	7	8	9	10	11	12	13	14
g^i	13−	11+	1−	7−	9+	15+	14+	4+	22+	10+	17+	8−	18+

Since g^{29-i} is the inverse of g^i, and the group is cyclic, this table is sufficient for all calculations in G_E. In particular, you will quickly see that my private key is $b' = 5$. Then, if you intercept some ciphertext intended for me, say

$$(9+, 15-) \quad (11+, 4-) \quad (0+, 18+) \quad (7-, 1+) \quad (14+, 4-)$$
$$(15+, 7+) \quad (13-, 18+) \quad (1-, 22+)$$

you can apply my decryption function $(\ell, c) \mapsto c * \ell^{-5}$ and obtain

$$14- \quad 7- \quad 4+ \quad I \quad 4+ \quad 10- \quad 1- \quad 22-$$

which is definitely

`THE⊔END.`

EXERCISES

15.21. The message above was encrypted using a different value of the token t for each symbol. Find these values.

15.22. Karen has agreed to use ElGamal cryptosystem based on the group G_E defined above, with the 'standard' representation of extended `english`. She has chosen the generator $h = 4-$, and her public key is $k = 9+$. I have sent her the message

$$(7-, 7+) \quad (8-, 9+) \quad (18-, 4-) \quad (10+, 0+) \quad (1-, 8-)$$
$$(15-, 7+) \quad (8-, 18-) \quad (17-, 10+) \quad (7+, 4-) \quad (8-, 4-) \quad .$$

What does it say?

Further reading for Chapter 15

There are several books on the mathematical theory of elliptic curves, at various levels of sophistication. From the cryptographic perspective there are two fundamental results: Hasse's theorem on the order of G_E (Section 15.4), and a theorem that says G_E can be expressed as the product of at most two cyclic groups. These results are discussed in the books by Silverman [**15.5**] and Washington [**15.6**], among others.

The rapidly developing field of elliptic curve cryptography is surveyed in two books by Blake, Seroussi, and Smart [**15.1**, **15.2**]. These books cover many of the implementation issues including the cost of exponentiation (Section 15.5). Further details on addition chains, the NAF, and so on, can be found in the survey by Gordon [**15.3**], and the famous tome by Knuth [**15.4**].

15.1 I. Blake, G. Seroussi, N. Smart. *Elliptic Curves in Cryptography.* Cambridge University Press (1999).

15.2 I. Blake, G. Seroussi, N. Smart. *Advances in Elliptic Curve Cryptography.* Cambridge University Press (2005).

15.3 D.M. Gordon. A survey of fast exponentiation algorithms. *J. Algorithms* 27 (1998) 129-146.

15.4 D.E. Knuth *The Art of Computer Programming II - Semi-numerical Algorithms.* Addison-Wesley (third edition, 1997).

15.5 J.H. Silverman. *The Arithmetic of Elliptic Curves.* Springer-Verlag (1986).

15.6 L.C. Washington *Elliptic Curves: Number Theory and Cryptography.* Chapman and Hall / CRC Press (2003).

Answers to odd-numbered exercises

Answers to the odd-numbered exercises are given below. In some cases the 'answer' is just a hint, in others there is a full discussion. A complete set of answers to all the exercises, password-protected, is available to instructors via the Springer website. To apply for a password, visit the book webpage at www.springer.com or email textbooks@springer.com.

Chapter 1

1.1. 'Canine' has six letters and ends in 'nine'. The second message has two possible interpretations.

1.3. The *mathematical bold* symbols $\mathbb{A}$ and $\mathbb{B}$.

1.5. This exercise illustrates the point that decoding a message requires the making and testing of hypotheses. Here the rules are fairly simple, but that is not always so. In the first example, it is a fair guess that the numbers represent letters, and the simplest way of doing that is to let $1, 2, 3, \ldots$ represent `A`, `B`, `C`, ... The number 27 probably represents a space. Testing this hypothesis, we find the message `GOOD␣LUCK`. The second example has the same number of symbols as the first, and each is represented by a word with 5 bits. How is this word related to the corresponding number in the first example?

1.7. s_1s_2 and s_3s_1 are both coded as 10010.

1.9. In both cases S is the 27-symbol alphabet $\mathbb{A}$. In the first example $T = \{1, 2, \ldots, 27\}$, and the coding function uses only strings of length 1. In the second example $T = \mathbb{B}$, and the coding function $S \to \mathbb{B}^*$ is an injection into the subset $\mathbb{B}^5$ of $\mathbb{B}^*$.

1.11. SOS; MAYDAY.

1.13. The number of ways of choosing 2 positions out of 8 is the binomial number $\binom{8}{2} = (8 \times 7)/2 = 28$. Hence at most 28 symbols can be represented in the semaphore code.

1.15. Using words of length 2 there are only 4 possible codewords, so we need words of length 3, where we can choose any 6 of the 8 possibilities, say

$$1 \mapsto 001 \quad 2 \mapsto 010 \quad 3 \mapsto 011 \quad 4 \mapsto 100 \quad 5 \mapsto 101 \quad 6 \mapsto 110.$$

With this code, if one bit in a codeword is wrong, then the result is likely to be another codeword: for example, if the first bit in 110 is wrong, we get the codeword 010. This problem cannot be avoided if we are restricted to using words of length 3. In order to overcome the problem we must use codewords with the property that any two differ in at least two bits. In that case, if one bit in any codeword is in error, then the result is not a codeword, and the error will be detected. This can be arranged if we use codewords of length 4, for example

$$1 \mapsto 0000 \quad 2 \mapsto 1100 \quad 3 \mapsto 1010 \quad 4 \mapsto 1001 \quad 5 \mapsto 0110 \quad 6 \mapsto 0101.$$

1.17. No, because the message refers to `ENGLAND`, which did not exist in Caesar's time. Also it is written in English.

Chapter 2

2.1. $s_3 s_4 s_2 s_1 s_4 s_2 s_3 s_1$.

2.3. The new code is $s_1 \mapsto 10$, $s_2 \mapsto 1$, $s_3 \mapsto 100$. Clearly it is not prefix-free, since 1 is a prefix of both 10 and 100. However, it can be decoded uniquely by noting that each codeword has the form 1 followed by a string of 0's Alternatively decoding can be done by reading the codeword backwards. If the last bit is 1, the last symbol must be s_2. If it is 0, looking at the last-but-one bit enables us to decide if the last symbol is s_1 or s_3. Repeating the process the entire word can be decoded uniquely. For example, 110101100 decodes as $s_2 s_1 s_1 s_2 s_3$.

2.5. The code can be extended by adding words such as 011, 101, without losing the PF property.

2.7. 128.

2.9. (i) $00, 101, 011, 100, 101, 1100, 1101, 1110$; (ii) $0, 100, 101, 1100, 1101, 1110, 11110, 11111$.

2.11. In part (i) take $T = \{0, 1, 2\}$; then the codewords could be 00 and any 12 of the 18 words $1**$, $2**$.

2.13. The parameters n_1, n_2, n_3, n_4 must satisfy

$$n_1 + n_2 + n_3 + n_4 = 12, \quad \frac{n_1}{2} + \frac{n_2}{4} + \frac{n_3}{8} + \frac{n_4}{16} \leq 1.$$

These equations imply that $7n_1 + 3n_2 + n_3 \leq 4$, so $n_1 = 0$ and $n_2 \leq 1$. Now it is easy to make a list of the possibilities:

$$\begin{array}{cccccccc} n_2: & 1 & 1 & 0 & 0 & 0 & 0 & 0 \\ n_3: & 1 & 0 & 4 & 3 & 2 & 1 & 0 \\ n_4: & 10 & 11 & 8 & 9 & 10 & 11 & 12 \end{array}.$$

2.15. The coefficient of x^4 in $Q_2(x)$ is the number of S-words of length 2 that are represented by T-words of length 4. These 25 words are all the words $s_i s_j$ with $i, j \in \{2, 3, 4, 5, 6\}$.

2.17. Use the fact that $Q_r(x) = Q_1(x)^r$.

Chapter 3

3.1. Occurrences of: a, about 60; ab, about 18.

3.3. This is a *deterministic* source. The probability distribution associated with each ξ_k is trivial. For example, $\Pr(\xi_4 = 3) = 1$, $\Pr(\xi_4 = n) = 0$ for $n \neq 3$.

3.5. Suppose the word-lengths are $x_1 \leq x_2 \leq x_3$, so the average is $L = x_1\alpha + x_2\beta + x_3(1 - \alpha - \beta)$. The KM condition is

$$\frac{1}{2^{x_1}} + \frac{1}{2^{x_2}} + \frac{1}{2^{x_3}} \leq 1.$$

The least possible value of x_1 is 1. If also $x_2 = 1$ there is no possible value for x_3. If $x_2 = 2$, we must have $x_3 = 2$. So we get the 'obvious' solution $x_1^* = 1, x_2^* = 2, x_3^* = 2$, which gives $L = 2 - \alpha$. Any other solution must have $x_i \geq x_i^*$, with at least one strict inequality, and since L is an increasing function of x_1, x_2, x_3, the average word-length will also be greater.

3.7. 0.

3.9. Start by proving that

$$UH(u_1/U, u_2/U, \ldots, u_m/U) = U \log U + \sum_{i=1}^{m} u_i \log(1/u_i).$$

3.11. $h'(x) = (1/\ln 2)\log((1-x)/x)$. This is zero when $(1-x)/x = 1$, that is, $x = \frac{1}{2}$. As x tends to 0 or 1, $\log((1-x)/x)$ tends to $\pm\infty$.

3.13. The SF rule says that the word-lengths are such that x_1 is the least integer such that $2^{x_1} \geq (1/0.25) = 4$, that is, $x_1 = 2$, and so on. The results are $x_1 = 2$, $x_2 = 4$, $x_3 = 3$, $x_4 = 5$, $x_5 = 3$, $x_6 = 2$. The average word-length is 2.7, and the entropy is $H(\mathbf{p}) \approx 2.42$.

3.15. The probabilities at each stage are as follows (no attempt has been made to make them increase from left to right). The new entry in each line is in bold.

0.25		0.1			0.15	0.05	0.2		0.25
0.25			**0.15**		0.15		0.2		0.25
0.25				**0.3**			0.2		0.25
0.25				0.3				**0.45**	
	0.55							0.45	
					1.0				

Using rule **H2** gives the following codewords (minor variations are possible): 01, 0011, 000, 0010, 10, 11, with $L = 2.45$.

3.17. The entropy is 2.72 approximately. The SF rule gives codewords of lengths $3, 3, 3, 4, 4, 4, 4$ with average word length $L_{SF} = 3.4$. This satisfies the condition $H(\mathbf{p}) \leq L_{SF} < H(\mathbf{p}) + 1$.

The sequence of probability distributions generated by the Huffman rule is as follows (the probabilities have not been re-ordered on each line).

0.2	0.2	0.2	0.1	0.1	0.1	0.1
0.2	0.2	0.2	0.1	0.1	0.2	
0.2	0.2	0.2	0.2		0.2	
0.2	0.2	0.2	0.4			
0.2	0.4		0.4			
0.6			0.4			
1.0						

Using **H2** the codewords are 00, 010, 011, 100, 101, 110, 111. (Several choices are possible, but all of them will produce a code with one word of length 2 and six of length 3.) The average word-length is $L_{opt} = 2.8$.

3.19. Use Lemma 3.17.

3.21. Using tree diagrams a few trials produces the code 0, 10, 11, 12, 20, 21, 22, which has average word-length 1.8. Since the probabilities are all multiples of $1/10$, the average word length of any such code is a number of the form $m/10$. However the entropy with respect to encoding by a ternary alphabet is $H_3(\mathbf{p}) = H(\mathbf{p})/\log_2 3 \approx 1.72$. Hence the word-length 1.8 must be optimal.

Chapter 4

4.1. $k \geq r^\ell + \ell - 1$.

4.3. The entropy of the given distribution is approximately 0.72. For the obvious code $A \mapsto 0$, $B \mapsto 1$, clearly $L_1 = 1$. For blocks of length 2 the probabilities are $0.64, 0.16, 0.16, 0.04$, and a Huffman code is $AA \mapsto 0$, $AB \mapsto 10$, $BA \mapsto 110$, $BB \mapsto 111$. This has average word-length 1.56 and $L_2/2 = 0.78$. For blocks of length 3 the probabilities and a Huffman code are

AAA	AAB	ABA	BAA	ABB	BAB	BBA	BBB
0.512	0.128	0.128	0.128	0.032	0.032	0.032	0.008
0	100	101	110	11100	11101	11110	11111.

Thus $L_3 = 2.184$ and $L_3/3 = 0.728$. This suggests that the limit of L_n/n as $n \to \infty$ is the entropy, 0.72 approximately.

4.5. $H(\mathbf{p}) \approx 2.446$, $H(\mathbf{p}') \approx 0.971$, $H(\mathbf{p}'') \approx 1.571$.

4.7. $\mathbf{p}^1 = [0.4, 0.3, 0.2, 0.1]$. Not memoryless.

4.9. $H \leq H(\mathbf{p}^2)/2 \approx 1.26$.

4.11. Use the hint given.

4.13. It is sufficient to consider the range $0 < x < 0.5$, when the original distribution is $[x^2, x(1-x), x(1-x), (1-x)^2]$ and the numerical values increase the order given. The first step is to amalgamate x^2 and $x(1-x)$, giving the distribution $[x, x(1-x), (1-x)^2]$. Since $x > x - x^2$ the middle term is always one of the two smallest. The other one is x if $0 < x \leq q$ and $(1-x)^2$ if $q < x \leq 0.5$, where q is the point where $x = (1-x)^2$. In fact, $q = (3-\sqrt{5})/2 \approx 0.382$. In the first case the word-lengths of the optimal code are $3, 3, 2, 1$, and in the second case they are $2, 2, 2, 2$. Hence

$$L_2(x) = 1 + 3x - x^2 \quad (0 < x \leq q), \qquad L_2(x) = 2 \quad (q < x \leq 0.5).$$

4.15. In binary notation $1/3$ is represented as $.010101\ldots$, where the sequence 01 repeats for ever. In general a rational number is represented by an expansion that either terminates or repeats, depending on the base that is used. For example, the representation of $1/3$ repeats in base 2 and base 10, but terminates in base 3.

4.17. It is enough to calculate the values of n_P, since the average word-length is $L_2 = 1 + \sum P n_P \approx 4.44$. Since $H(\mathbf{p}^1) \approx 1.685$; $H(\mathbf{p}^2)/2 \approx 1.495$, the entropy does not exceed 1.495, whereas $L_2/2 \approx 2.22$.

4.19. For $X = x_1 x_2 \ldots x_{r-1} x_r$ let $X^* = x_1 x_2 \ldots x_{r-1} \alpha$, where α is the first element of S. The probabilities $P(Y)$ with $Y < X$ can be divided into two sets: those with $Y < X^*$ and those with $X^* \leq Y < X$.

4.21. $L_2 \approx 5.01$. $H = H(P) \approx 1.68 < L_2/2$.

4.23. Check that the encoding rule creates dictionaries in Example 4.25.

4.25. The message begins BET␣ON␣TEN␣...

Chapter 5

5.1.

$$\begin{pmatrix} 1-a & a \\ b & 1-b \end{pmatrix}.$$

5.3. Let $\mathbf{p}$ be the input to Γ_1, $\mathbf{q}$ the output from Γ_1 and input to Γ_2, $\mathbf{r}$ the output from Γ_2. Then $\mathbf{r} = \mathbf{q}\Gamma_2 = \mathbf{p}\Gamma_1\Gamma_2$. Hence Γ is the matrix product $\Gamma_1\Gamma_2$.

5.5. Since $e_{n+1} = e_n + e - 2ee_n$ it follows that $e_n \to \frac{1}{2}$ as $n \to \infty$. When $e = 0$, we have $e_n = 0$ for all n, so the limit is zero. When $e = 1$ the value of e_n alternates between 0 and 1, so there is no limit.

5.7. $q_0 = 0.98p_0 + 0.04p_1$, $q_1 = 0.02p_0 + 0.96p_1$.

5.9. $t_{00} = 0.594$, $t_{01} = 0.006$, $t_{10} = 0.004$, $t_{11} = 0.396$. $H(\mathbf{t}) \approx 1.0517$ and (as in Exercise 5.6) $H(\mathbf{q}) \approx 0.9721$, hence $H(\Gamma;\mathbf{p}) \approx 0.0796$.

5.11. We have $\mathbf{q} = \mathbf{p}\Gamma = [pc\ (1-p)c\ 1-c]$, from which it follows by direct calculation that $H(\mathbf{q}) = h(c) + ch(p)$. For the joint distribution $\mathbf{t}$ the values are $pc, 0, p(1-c)$ and $0, (1-p)c, (1-p)(1-c)$. Thus (by the argument used the proof of Theorem 5.9) $H(\mathbf{t}) = h(p) + h(c)$, and

$$H(\Gamma;\mathbf{p}) \;=\; H(\mathbf{t}) - H(\mathbf{q}) = h(p) + h(c) - h(c) - ch(p) = (1-c)h(p).$$

5.13. (i) 1/27; (ii) 1; (iii) 3.

5.15. The channel matrix Γ is a $2N \times 2$ matrix with rows alternately 0 1 and 1 0. If the input source is $[p_1, p_2, \ldots, p_{2N}]$, the joint distribution $\mathbf{t}$ is given by

$$t_{1,0} = 0,\ t_{1,1} = p_1,\ t_{2,0} = p_2,\ t_{2,1} = 0,\quad \ldots\quad,\ t_{2N,0} = p_{2N},\ t_{2N,1} = 0.$$

Hence $H(\mathbf{t}) = H(\mathbf{p})$. It follows that

$$H(\mathbf{p}) - H(\Gamma;\mathbf{p}) \;=\; H(\mathbf{p}) - H(\mathbf{p} \mid \mathbf{q}) \;=\; H(\mathbf{p}) - H(\mathbf{t}) + H(\mathbf{q}) \;=\; H(\mathbf{q}).$$

So the maximum of $H(\mathbf{p}) - H(\Gamma;\mathbf{p})$ is the maximum of $H(\mathbf{q})$, which is 1. This means that the channel can transmit one bit of information about any input: specifically, it tells the Receiver whether the input is odd or even.

5.17. For each input x,

$$H(\mathbf{q} \mid x) = 2\alpha \log(1/\alpha) + (1-2\alpha)\log(1/(1-2\alpha)) = k(\alpha) \text{ say.}$$

Hence $H(\mathbf{q} \mid \mathbf{p}) = k(\alpha)$, which is constant. The maximum of $H(\mathbf{q})$ occurs when $\mathbf{q} = [1/4, 1/4, 1/4, 1/4]$, and so the capacity is $2 - k(\alpha)$.

Chapter 6

6.1. One of the instructions S,E,W.

6.3. Yes, because 100100 is more like 000000 than any other codeword.

6.5. The first column is $(1-a)^2$, $(1-a)b$, $(1-a)b$, b^2.

6.7. c_7, c_5, c_3.

6.9. The nearest codewords for z_1 are $11000, 01100, 01010, 01001$. So $\sigma(z_1)$ is certainly not 10001, and this event has probability zero. The nearest codewords for z_2 are $11000, 10100, 10010, 10001$. The rule is that one of them is chosen as $\sigma(z_2)$ with probability $1/4$, so the probability that z_2 is received and $\sigma(z_2) = 10001$ is $e/4$. The nearest codewords for z_3 and z_4 do not include 10001, so (like z_1) these contribute nothing. The nearest codewords for z_5 are $01001, 10001, 11000$. The rule is that one of them is chosen as $\sigma(z_5)$ with probability $1/3$, so the probability that z_5 is received and $\sigma(z_5) = 10001$ is $e/3$. Hence the required probability is $7e/12$.

6.11. The MD rule would give $\sigma(100) = 000$, for example, which is clearly not the same as the maximum likelihood rule in this case.

6.13. For each $c \in C$, there are 6 words that can result from one bit-error, so $|N_1(c)| = 7$. Any of the resulting $5 \times 7 = 35$ words can be corrected, so the number of words that cannot be corrected is $64 - 35 = 29$.

6.15. We require $\rho = (\log_2 |C|)/6$ to be at least 0.35. Thus $\log_2 |C| \geq 2.1$, which means that $|C| \geq 5$. For an example, see Exercise 6.12.

6.17. $n = 14$.

6.19. Suppose C is a maximal code for the given values of n and r. Then the neighbourhoods $N_{2r}(c)$ $(c \in C)$ must completely cover $\mathbb{B}^n$, otherwise there would be a word $x \in \mathbb{B}^n$ such that $d(x, c) \geq 2r + 1$ for all $c \in C$, contradicting the maximality of C.

Chapter 7

7.1. For each $c \in C$ there are two other codewords. The MD rule does not assign these codewords to c, nor the words at distance 1 from them. Hence $F(c)$ contains at least $2 \times (1 + 5) = 12$ words.

7.3. If $\mathbf{p} = (0.9, 0.1)$ we have $\mathbf{q} = (0.74, 0.26)$. For the ideal observer rule we need the conditional probabilities $\Pr(c \mid z)$ which can be calculated using the rule $\Pr(c \mid z)q_z = \Pr(z \mid c)p_c$, where $\Pr(z \mid c) = \Gamma_{cz}$. The numbers are $\Pr(c = 0 \mid z = 0) \approx 0.98$, $\Pr(c = 0 \mid z = 1) \approx 0.70$, $\Pr(c = 1 \mid z = 0) \approx 0.02$, $\Pr(c = 1 \mid z = 1) \approx 0.30$. The ideal observer rule σ^* says that $\sigma^*(z) = c$ when $\Pr(c \mid z) \geq \Pr(c' \mid z)$ for all $c' \in C$. In this case $\sigma^*(0) = 0$, $\sigma^*(1) = 0$: in other words, the Receiver always decides that 0 was the intended codeword. Hence $M_0 = 0$, $M_1 = 1$, and the probability of a mistake is $0.9M_0 + 0.1M_1 = 0.1$.

7.5. $10e^3(1-e)^2 + 5e^4(1-e) + e^5 = 10e^3 - 15e^4 + 6e^5$.

7.7. Since $|R_n| = 2$, the information rate is $1/n$, which tends to 0 as $n \to \infty$. As in Exercise 6.10, the channel matrix has the form

$$\begin{array}{lcccccc} 00\cdots0 & 1 & 0 & 0 & . & . & . & 0 \\ 11\cdots1 & f^n & gf^{n-1} & gf^{n-1} & . & . & . & g^n \end{array},$$

where $g = 1-f$. Thus the maximum likelihood rule gives $\sigma(00\cdots0) = 00\cdots0$, but $\sigma(z) = 11\cdots1$ for all $z \neq 11\cdots1$. A mistake occurs only when $11\cdots1$ is sent and all bits are transmitted wrongly. The probability of a mistake is therefore $(1-p)f^n$, which also tends to 0.

7.9. The code in Example 7.5 has parameters $(6,3,3)$, and the new code has parameters $(6,3,2)$.

7.11. Since $e = 0.03$, we have $\gamma = 1 - h(e) \approx 0.80$. Thus the uncertainty is at least $0.1\,n$, where n is the length of the codewords.

7.13. Use the fact that $H(\mathbf{q}) \leq H(\mathbf{q}') + H(\mathbf{q}'')$.

7.15. Let $\mathbf{p} = [p, 1-p]$. Since $e = 0.5$, $\mathbf{q} = [0.5, 0.5]$. Hence the left-hand side of Fano's inequality is $h(p) - h(e) + h(q) = h(p)$. Since $M = e = 0.5$ (Example 7.2) the right-hand side is 1, and the inequality is just $h(p) \leq 1$. In particular, when $p = 0.5$ there is equality.

Chapter 8

8.1. C_1 is linear but C_2 is not, because (for example) $100 + 010 = 110$ which is not in C_2.

8.3. (a) A linear code with dimension k has 2^k codewords, so if there are 53 students we require $2^k \geq 53$. Thus $k = 6$ is the minimum possible value.

(b) Suppose that the codewords have length n, and the code allows for the correction of one error. Then the neighbourhoods $N_1(c)$, consisting of a codeword c and the n words that can be obtained from c by changing one bit, must be disjoint. There are 2^6 neighbourhoods, so $2^6(1+n) \leq 2^n$, that is, $n + 1 \leq 2^{n-6}$. The least value of n for which this is true is 10.

8.5. Since $0+0 = 0$ and $1+1 = 0$, the weight of $x+y$ is equal to the number of places where x and y differ. This is equal to

$$\begin{array}{l}(\text{number of places where } x \text{ is 1 and } y \text{ is 0}) \\ \quad + (\text{number of places where } x \text{ is 0 and } y \text{ is 1}).\end{array}$$

The first term is equal to $w(x) - w(x*y)$ and the second term is equal to $w(y) - w(x*y)$. Hence $w(x+y) = w(x) + w(y) - 2w(x*y)$. In particular, if x and y both have even weight, so does $x+y$. This proves that the set of words of even weight is a linear code.

8.7. Let I denote the $k \times k$ identity matrix. Then

$$E = \begin{pmatrix} & & & I & & & \\ 1 & 1 & . & . & . & 1 & 1 \end{pmatrix}.$$

8.9. $\rho = k/(k+1)$, $\delta = 2$.

8.11. The matrix $[1\ 1\ 1\ \ldots\ 1]$, with $k+1$ columns.

8.13. A suitable matrix is given in Exercise 8.17.

8.15. It helps to note that columns 2,3,5,6 are the columns of the identity matrix (in scrambled order). This means that x_2, x_3, x_5, x_6 are determined by x_1, x_4, x_7. Explicitly, we can write the equations in the form

$$\begin{aligned} x_2 &= x_1 + x_4 + x_7 \\ x_5 &= x_4 \qquad + x_7 \\ x_3 &= x_1 + x_4 + x_7 \\ x_6 &= \qquad\qquad x_7. \end{aligned}$$

Thus there are 8 codewords, corresponding to the 2^3 choices for the bits x_1, x_4, x_7. For example, if $x_1 = 0, x_4 = 0, x_7 = 1$ the equations say that $x_2 = 1$, $x_3 = 1$, $x_5 = 1$ and $x_6 = 1$. Since the columns of H are non-zero and all different, Theorem 8.10 implies that the minimum distance is 3, at least. Thus $n = 7$, $k = 3$, $\delta = 3$.

8.17. Given $z' = [111010]'$, we find that $Hz' = [110]'$. This is the second column of H, so there was an error in the second bit, and the intended codeword was 101010.

8.19. The given check matrix defines x_3, x_4, x_5 in terms of x_1, x_2, by the equations $x_3 = x_1$, $x_4 = x_2$, $x_5 = x_1 + x_2$. The code words are obtained by giving the values 00, 01, 10, 11 to x_1, x_2 and using the equations. So they are 00000, 01011, 10101, 11110. There are eight cosets and the syndromes and coset leaders are:

000	100	010	001	110	101	011	111
00000	00100	00010	00001	11000	10000	01000	01100.

Suppose the received word is $z = 11111$. Then the syndrome Hz' is 001, which corresponds to the coset leader $f = 00001$. According to the rule, $\sigma(z) = z + f = 11110$, which is indeed a codeword. In the other cases the codewords are 11110, 10101, 11110.

8.21. The syndrome of z is $10000000'$. It is not the zero word, or a column of H, so more than one bit-error has occurred. If two bit-errors have occurred, the syndrome must be the sum of two columns of H. The possiblities for the first four rows are $(\mathbf{h}_1, \mathbf{h}_9)$, $(\mathbf{h}_2, \mathbf{h}_{10})$, and so on. The corresponding pairs in the bottom four rows are $(\mathbf{h}_5, \mathbf{h}_8)$, $(\mathbf{h}_3, \mathbf{h}_5)$, and so on. Checking these pairs we find the sixth pair is $(\mathbf{h}_{15}, \mathbf{h}_{15})$ and $\mathbf{h}_{15} + \mathbf{h}_{15} = 0000'$. This corresponds to $(\mathbf{h}_6, \mathbf{h}_{14})$, so errors have occurred in bits 6 and 14.

Chapter 9

9.1. The check matrix can be obtained by writing down the 15 columns corresponding to the binary representations of the numbers $1, 2, \ldots, 15$, in order. The number of codewords is $2^{15-4} = 2048$. Denote the given words by z_1, z_2, z_3. Then $Hz_1' = [1010]'$, $Hz_2' = [0111]'$, $Hz_3' = [1000]'$. Hence z_1 has an error in the 10th bit, z_2 has an error in the 7th bit, and z_3 has an error in the 8th bit.

9.3. The number of cosets is $2^n/|C| = 2^n/2^{n-m} = 2^m$, where m is the number of rows of a check matrix. For a Hamming code $n = 2^m - 1$. The syndromes of the $n+1$ given words are all distinct.

9.5. $1+x+x^2$ represents the word 111. The other codewords are obtained by multiplying $1+x+x^2$ by $f(x)$, where $f(x)$ is any polynomial with degree less than 3, and reducing mod x^3-1. It turns out that there are only two possibilities 0 and $1+x+x^2$ itself, so the code is $\{000, 111\}$. Since the codewords are defined by the equations $x_1 = x_2 = x_3$ a suitable check matrix is

$$H = \begin{pmatrix} 1 & 1 & 0 \\ 0 & 1 & 1 \end{pmatrix}.$$

9.7. Yes. The code defined by the ideal $\langle 1+x \rangle$ contains all words with weight 2 and hence all words with even weight.

9.9. Three of the 8 divisors, 1, x^7-1, and $1+x+x^3$, are discussed in Example 9.14. The divisor $1+x$ generates the code containing all words with even weight, and the divisor $1+x^2+x^3$ generates a code equivalent to the Hamming code. The divisor $(1+x)(1+x+x^3)$ corresponds to $h(x) = 1+x^2+x^3$, and the resulting check matrix defines a code with parameters $(7,3,3)$. The divisor $(1+x)(1+x^2+x^3)$ is similar. The divisor $(1+x+x^3)(1+x^2+x^3)$ corresponds to $h(x) = 1+x$, and the resulting check matrix defines a code with parameters $(7,1,7)$, in other words the repetition code $\{0000000, 1111111\}$.

9.11. Since the Hamming code with word-length $15 = 2^4-1$ has dimension $11 = 2^4-1-4$, the canonical generator $g(x)$ must have degree 4 (Theorem 9.13). Taking $g(x) = 1+x+x^4$, the complementary factor is

$$h(x) = 1+x+x^2+x^3+x^5+x^7+x^8+x^{11}.$$

According to the rule given in the text, the corresponding matrix H has 4 rows and 15 columns, all of which are different, so it must define a code equivalent to the Hamming code. (Does the same result hold if we choose one of the other factors of degree 4?)

9.13. The usual method of evaluating a determinant by expanding in terms of the first row gives the result $a^{33} - a^{31} + a^{25} - a^{30} + a^{26} - a^{23}$, which factorizes as claimed.

9.15. The packing bound requires $16(1+n+\frac{1}{2}n(n-1)) \leq 2^n$. The smallest integer for which this holds is $n = 10$.

9.17. The word 100101 01101 is a codeword but its cyclic shift 110010 10110 is not.

9.19. $x^3 - 1 = (1+x)(1+x++x^2)$. For $dd = 3$, the canonical generator is $m_1(x) = 1+x+x^2$. Now refer to Exercise 9.5.

9.21. The Hamming code has parameters $(15, 11, 3)$ so it corrects 1 error and has rate $11/15$. The BCH code has parameters $(15, 7, 5)$, so it corrects 2 errors, but has a lower rate $7/15$.

9.23. Let α be as in Example 9.22, so that $m_1(x) = 1+x+x^4$. Using the table of powers of α it can be verified that $m_3(x) = 1+x+x^2+x^3+x^4$, $m_5(x) = 1+x+x^2$, $m_7(x) = 1+x^3+x^4$. Thus the canonical generators for the cases $dd = 7$ and $dd = 9$ have degrees 10 and 14 respectively, and the dimensions of the BCH codes are 5 and 1. The codes for $dd = 11, 13, 15$ are subcodes of the code for $dd = 9$, and in fact all these codes are simply the repetition code with just two words.

Chapter 10

10.1. Use the addition formula, as in Example 4.7.

10.3. It is convenient to write a b c instead of $*$! ? . The string becomes *cbaacbbaaccbbaacbaac* . An estimate of the probabilities for the first-order approximation is obtained by counting the frequencies of the letters in the given string: $p_a = 8/20 = 0.4$, $p_b = 6/20 = 0.3$, $p_c = 6/20 = 0.3$. Hence the entropy is

$$0.4\log_3(1/0.4) + 0.3\log_3(1/0.3) + 0.3\log_3(1/0.3) \approx 0.991.$$

Similarly, we can estimate the probabilities of the digrams. Those with non-zero frequency are

$$p_{aa}=4/19,\ p_{ac}=4/19,\ p_{ba}=4/19,\ p_{bb}=2/19,\ p_{cb}=4/19,\ p_{cc}=1/19.$$

From this we can calculate that the uncertainty (in bits per *symbol*) is approximately 0.776. The reduction in uncertainty is due to the predictability of the language. In particular the word !**? occurs four times in the given string.

10.5. The symbols x and y both have probability 0.5 so $\mathcal{U}_1 = 1$. A simple calculation gives $\mathcal{U}_2 \approx 0.985$. Since $\mathcal{U}$ is the infimum of the values $\mathcal{U}_n$ it follows that $\mathcal{U} \leq \mathcal{U}_2 < 1$.

10.7. `LOTS␣OF␣PEOPLE␣SUPPORT␣MANCHESTER␣UNITED.`

10.9. The `textish` alphabet has more symbols than `english`, because numerals are often used, for example 4 replaces FOR. It is reasonable to suppose that `textish` has greater uncertainty than `english`, because there are fewer rules.

10.11. The key is 9. MATHEMATICS␣IS␣OFTEN␣USEFUL .

10.13. In cycle notation, the encryption key is

(␣)(AROJNHCPKDUTSQMGI)(BE)(FL)(V)(W) (X)(Y)(Z) .

The ciphertext is CRKKY EAOSCURY IORHHY.

10.15. An encryption key that is the same as the decryption key is a permutation σ such that $\sigma = \sigma^{-1}$. If there are no fixed letters (1-cycles), the cycle form of σ must consist entirely of 2-cycles.

Let T_n denote the number of permutations of a set of size $2n$ that have n 2-cycles. Then $T_n = (2n - 1)T_{n-1}$. Hence there are $T_{13} = 25 \times 23 \times \cdots \times 3 \times 1 = 7905853580625$ possible keys. Although this a large number in everyday terms, if Eve has a reasonable computer she could carry out an exhaustive search quite quickly.

Chapter 11

11.1. (ADOKG)(BECMI)(FRPLHTSQNJ). The key for decryption is therefore (AGKOD) (BIMCE)(FJNQSTHLPRF) .

11.3. THEINDEXOFCOINCIDENCEISAPOWERFULTOOL.

11.5. UW IL IC ER RL NL LG IZ NG OV TW CN IM KV FX VG FM MF ST .

11.7. Suppose the ciphertext digram is ab. *Case 1*: if a and b are in different rows and columns, then $ab \mapsto xy$, where a, b, x, y are the corners of a rectangle and a, x are in the same row.

11.9. $[18\ 2]'[3\ 2]'[12\ 22]'[17\ 17]'$.

11.11. The block AMPLE corresponds to $[1, 13, 16, 12, 5]'$, which is encrypted as $[17, 5, 2, 6, 13]'$, so the required pair is $(17, 5)$. Similar calculations for the other blocks result in different pairs.

11.13. The decryption function is $x = \gamma y + \delta$, where $\gamma = \alpha^{-1}$ and $\delta = -\beta\alpha^{-1}$. Since γ and δ can be calculated easily when α and β are known, we have a symmetric key system. If two plaintext-ciphertext pairs (x_1, y_1) and (x_2, y_2) are known, then the equations $y_1 = \alpha x_1 + \beta$ and $y_2 = \alpha x_2 + \beta$ can be solved for α and β. (There is an exception – what is it?) If the plaintexts are chosen to be $x_1 = 0$ and $x_2 = 1$, the solution is especially simple.

Chapter 12

12.1. $H(\mathbf{r}) = 1$, $H(\mathbf{q}) = 2$, so $H(\mathbf{r} \mid \mathbf{q})$ is equal to $H(\mathbf{p}) - 1 \approx 0.846439$.

12.3. It is fairly easy to find two meaningful words of the form xyx that are obtained from one another by a cyclic shift: for example `DID` and `PUP`. A ciphertext aba that encrypts one of these words also encrypts the other, with a different key.

12.5. The columns of Γ are not constant (see Theorem 12.6).

12.7. `ELVIS␣LIVES`.

12.9. Suppose Eve knows $m_1 + m_2$ and conjectures that a segment y_1 of m_1 represents `TUESDAY`. If z is the corresponding segment of $m_1 + m_2$ the corresponding segment of m_2 is $y_2 = y_1 + z$. If y_2 is meaningful, it is likely that Eve's conjecture is correct.

12.11. Use the decryption rule given in Theorem 12.10.

12.13. $x' = 1100$, $y' = 1111$, so $d(x', y') = 2$.

12.15. The permutations are δ = (`ADOKG`)(`BECMI`) (`FRPLHTSQNJ`) and ρ = (`AROJNHCPKDUTSQMGI`)(`BE`)(`FL`). The double-locking procedure involves the application of the permutation $\rho^{-1}\delta^{-1}\rho\delta$ and it is easy to verify that it is not the identity permutation: for example, $\rho^{-1}\delta^{-1}\rho\delta$ (`A`) = `D`.

Chapter 13

13.1. The number of invertible elements is 12; $5^{-1} = 29$.

13.3. If $xy = 1 \pmod n$, say $xy = 1 + nk$, then

$$(n - x)(n - y) = n^2 - n(x + y) + xy = 1 + n(n - (x + y) + k),$$

so $n - y$ is the inverse of $n - x$. Hence the invertible elements occur in pairs, x and $n - x$ (which must be distinct). So the number is even.

13.5. Here $3599 = 61 \times 59$, so $\phi(n) = 60 \times 58 = 3480$. Thus the private key is given by $31d = 1 \pmod{3480}$, which is true when $d = 3031$.

13.7. By the Euclidean algorithm, $\gcd(15, 68) = 1$ and $15^{-1} = 59$.

13.9. The exponent has 10 decimal digits, so the crude estimate in the text gives an upper bound of 66. In fact the binary representation of 3578674567 has 31 binary digits, 18 of which 1's, so the actual number is $31 + 17 = 48$.

13.11. Using the fact that $1000 = 512 + 256 + 128 + 64 + 32 + 8$, it turns out that $2^{1000} = 27 \pmod{47}$.

13.13. $\phi(15) = 8$, so $d = 3$. It is easy to check that $m^9 = m \pmod{15}$ for $1 \leq m \leq 14$ (only the primes need be checked).

13.15. If $n = pq$ and $\gcd(m,n) \neq 1$ then (without loss of generality) we can take m to be a multiple of p, and trivially $m^{K+1} = m \pmod p$. Also $\gcd(m,q) = 1$ so $m^{q-1} = 1 \pmod q$. Hence

$$m^{K+1} = m^{k\phi(n)+1} = m^{k(p-1)(q-1)} \times m = m \pmod q.$$

So if $de = 1 \pmod \phi$, $m^{de} = m + Aq$. Since both m and m^{de} are multiples of p, so is A. Hence $m^{de} = m \pmod{pq}$.

13.17. Given $n = 4189$ and $\phi = 4060$, it follows that $p + q = n - \phi + 1 = 130$. Hence p and q are the roots of $x^2 - 130x + 4189 = 0$. Using the formula, we find $p = 71, q = 59$. Finally, $d = e^{-1} \pmod{4189} = 3023$.

13.19. The fact that n is prime can be established relatively easily. If that is known, then $\phi(n) = n - 1$ and the private key d is the inverse of e mod $n - 1$, which can be found easily using the Euclidean algorithm. Hence if $c = m^e$ is an encrypted message addressed to the user, it can be decrypted easily by the formula $m = c^d$.

Chapter 14

14.1. For each set of μ_y elements with hash value y there are $\mu_y(\mu_y - 1)/2$ collisions.

14.3. If h_k is not collision-resistant then given one pair $(m_1, h_k(m_1))$ it is easy to find m' with $h_k(m') = h_k(m)$, so $h_k(m')$ is known.

14.5. We have to show that the order of 3 is 16, in other words, the order is not 2, 4, or 8. We find $3^2 = 9$, $3^4 = 13$, $3^8 = 16$. In this case, the other primitive roots are the odd powers of 3.

14.7. We require $D_{a'}(E_a(m)) = m$, that is $(m^a)^{a'} = m$. This means $aa' = 1$ mod $p-1$, so $\gcd(a, p-1) = 1$. Similarly, $\gcd(b, p-1) = 1$. The double-locking procedure works in this case because the functions commute.

14.9. The following table shows n and 2^n mod 19.

1	2	3	4	5	6	7	8	9
2	4	8	16	13	7	14	9	18

10	11	12	13	14	15	16	17	18
17	15	11	3	6	12	5	10	1

2 is primitive root, because it has order 18. Since $2^{15} = 12$ the logarithm of 12 is 15, and since $2^{11} = 15$ the logarithm of 15 is 11.

14.11. The result follows from the equation $r^{\alpha+\beta} = r^\alpha r^\beta$, remembering that α and β are integers in the range $1 \leq \alpha, \beta \leq p-1$ and $r^{p-1} = 1$.

14.13. $(\ell, c) = (r^t, xk^t) = (7777, 6532)$.

14.15. Since A's private key is $x_A = 33$, her public key is $y_A = 2^{x_A} = 2^{33}$ (mod 149). If necessary, this can be calculated by hand: $2^2 = 4$, $2^4 = 16$, $2^8 = 256 = 107$, $2^{16} = 107^2 = 125$, $2^{32} = 125^2 = 129$, $2^{33} = 109$.

14.17. Applying the test given in Theorem 14.13, we find

$$a^{|i|}i^j = 6^2 \times 2^4 = 1, \qquad r^{|m|} = 5^6 = 8,$$

so the signature is not valid.

14.19. The verification test is $a^{|i|}i^j = r^{|m|}$, where a, r, m are known. If i is given a specific value then i^j is known, and finding j is a DLP in $\mathbb{F}_p^\times$. On the other hand if j is given a specific value, i occurs twice in the resulting equation, and the problem is (apparently) more difficult.

Chapter 15

15.1. The first four numbers have orders $1, 2, 4, 4$ respectively, and so they are not generators. All the other numbers are generators. (If you know that the numbers are the eighth roots of unity in $\mathbb{C}$, this is trivial.)

15.3. For any n, 1 is a generator, and $1 + 1 + \cdots + 1$ (i times) is i. Hence, given h, the solution to the DLP $1 + 1 + \cdots + 1$ (i times) $= h$ is $i = h$.

15.5. Alice chooses t and sends $(3^t, m \times 13^t)$. Bob's private key b' is the solution of the DLP $3^{b'} = 13$ in $\mathbb{F}_{17}^\times$, and by trial and error $b' = 4$. So Bob decrypts using the rule $(m \times 13^t) \times ((3^t)^{-1})^4$, which reduces to m, for all t.

15.7. $(1, 6) * (1, 6) = (0, 0)$ (Example 15.7), and $(0, 0) * (0, 0) = I$, by case (i) of Theorem 15.6. So $(1, 6)$ has order 4.

15.9. For any x, case (i) of Theorem 15.6 says that $(x, 0) * (x, 0) = I$. Conversely, if $y \neq 0$ then $(x, y) * (x, y)$ is determined by case (ii) and is not I.

15.11. Let $f = (2, 1)$. Then $f^2 = (0, 3)$, $f^3 = (4, 1)$, $f^4 = (4, 4)$, $f^5 = (0, 2)$, $f^6 = (2, 4)$, $f^7 = I$.

15.13. When $\beta = 6$ the relevant numbers are $m_1 = 0$ (by the previous exercise) and $m_2 = 6$ (by explicit calculation). Hence the group has order 13, which is prime, and so it is cyclic. Any element except I is a generator; for example $(-1, 2)$.

15.15. One element of order 5 is $(4, 4)$ (there are others).

15.17. The addition-subtraction chain $1, 2, 4, 8, 16, 32, 64, 128, 127$ has length 8.

15.19. If the binary representation has two adjacent 1's, say $c_{i-1} = c_i = 1$, the identity $2^i = 2^{i+1} - 2 \times 2^{i-1}$ gives a new representation with $c'_{i-1} = -1$, and $c'_i = 0$. The result follows by applying this transformation recursively.

15.21. The values of t are: 6 3 1 5 8 7 2 4.

Index

addition chain, 248
addition-subtraction chain, 249
Advanced Encryption Standard, 203
AES, 203
alphabet, 4
arithmetic code, 63
ASCII, 3
authentication, 221
average word-length, 30

b-ary, 5
BCH code, 156
binary, 5
binary asymmetric channel, 76
binary erasure channel, 80
binary symmetric channel, 74
bit, 2
bit-error probability, 74
BSC, 74

Caesar's system, 170
canonical generator, 149
capacity, 82
channel, 74
channel matrix, 74
characteristic 2, 241
check bits, 130
check matrix, 129
chosen plaintext attack, 187
ciphertext, 172
code, 5
codeword, 5
collision, 223
collision-resistant, 223
concatentation, 6
conditional entropy, 79
coset, 136
coset leader, 137
cryptosystem, 179
cumulative probability function, 63
cyclic code, 145
cyclic group, 237
cyclic shift, 145

Data Encryption Standard, 202
decision rule, 91
decryption functions, 172
DES, 202
designed distance, 156
dictionary, 67
dictionary order, 62
Diffie-Hellman system, 230
digram, 164
dimension, 124
Discrete Logarithm Problem, 226
DLP, 226
double-locking, 204

ElGamal cryptosystem, 228
ElGamal signature scheme, 234
elliptic curve, 241
encoded stream, 13, 90
encryption, 171
encryption functions, 171
english, 163
entropy, 32

entropy of stationary source, 55
error-correcting code, 101
Euclidean algorithm, 212
exhaustive search, 173
extended BSC, 94
extended channel, 94

Fano's inequality, 117
Feistel iteration, 199
final stream, 91
frequency analysis, 175
frequency table, 164

generator (of a cyclic group), 237
Gilbert-Varshamov bound, 105

Hamming code, 141
Hamming distance, 95
hash function, 222
Hasse's theorem, 247
Hill's system, 186
Huffman's rule, 40

ideal, 147
ideal observer rule, 96
independent, 49
index, 67
index of coincidence, 182
information, 35
information rate, 104
integrity, 221
inverse probabilities, 87

key, 171
key distribution problem, 204
key equivocation, 193
keyword, 175
known ciphertext attack, 187
known plaintext attack, 187
Kraft-McMillan number, 18

left-inverse, 172
length, 4
linear code, 124
LZW encoding, 68

MAC, 223
marginal distributions, 49
maximum likelihood rule, 97
MD rule, 97
memoryless source, 29
message, 4
message authentication code, 223
message bits, 130
minimum distance, 100
minimum distance rule, 97
mistake, 91
mono-alphabetic substitution, 174
Morse code, 7

NAF, 250
neighbourhood, 100
noisy channel, 74
non-adjacent form, 250
non-repudiation, 221

one-time pad, 197
optimal code, 31
original stream, 13, 90

packing bound, 102
parity check, 128
password, 227
perfect code, 143
perfect secrecy, 195
PF, 15
phi function, 208
plaintext, 172
Playfair system, 183
point at infinity, 241
poly-alphabetic encryption, 180
prefix-free, 15
primality testing, 211
primitive, 155
primitive root, 225
private key, 207
probability distribution, 28
probability of a mistake, 108
product channel, 94
public key, 207
public key cryptography, 207

rate, 104
received stream, 90
redundancy, 169
repeated-squaring, 213
root, 16
RSA cryptosystem, 207

S-box, 202
semaphore, 8
SF rule, 39
Shannon's theorem, 119
Shannon-Fano rule, 39
side-channel attacks, 220
signature, 232
signature functions, 232
signature scheme, 232

source, 27
spurious key, 194
standard form, 130
stationary source, 53
stream, 13
string, 4
subspace, 123
symmetric key cryptosystem, 179
syndrome, 135
syndrome look-up table, 137

ternary, 5
ternary asymmetric channel, 76
triangle inequality, 96

UD, 6
uncertainty, 34
uncertainty of a natural language, 166
unicity point, 194
Unicode, 3
uniquely decodable, 6

valid signature, 232
verification algorithm, 232
Vigenère system, 180

weight, 124
word, 4

Printed in the United States